Supply Chain Management for Bioenergy and Bioresources

Supply Chain Management for Bioenergy and Bioresources

Special Issue Editors

Dionysis Bochtis
Charisios Achillas

MDPI • Basel • Beijing • Wuhan • Barcelona • Belgrade • Manchester • Tokyo • Cluj • Tianjin

Special Issue Editors
Dionysis Bochtis
Institute for Bio-Economy and Agri-Technology (iBO)
Greece

Charisios Achillas
Technical Educational Institute of Central Macedonia
Greece

Editorial Office
MDPI
St. Alban-Anlage 66
4052 Basel, Switzerland

This is a reprint of articles from the Special Issue published online in the open access journal *Energies* (ISSN 1996-1073) (available at: https://www.mdpi.com/journal/energies/special_issues/Bioenergy_Biorecourses).

For citation purposes, cite each article independently as indicated on the article page online and as indicated below:

LastName, A.A.; LastName, B.B.; LastName, C.C. Article Title. *Journal Name* **Year**, *Article Number*, Page Range.

ISBN 978-3-03943-462-6 (Pbk)
ISBN 978-3-03943-463-3 (PDF)

Contents

About the Special Issue Editors

Dionysis Bochtis works on the area of systems engineering focused on bio-production and related provision systems with enhanced ICT and automation technologies up to fully robotized production systems. Throughout his carrier he has held various positions, including Professor in Agri-Robotics at the Lincoln Institute for Agri-Food Technologies, University of Lincoln, UK, and Senior Scientist in Operations Management at the Department of Engineering of Aarhus University, Denmark. Currently, he is the Director of the Institute for Bio-Economy and Agri-Technology (IBO), Centre for Research & Technology Hellas (CERTH). He runs numerous research projects on ICT and robotics in agricultural production. He is author of more than 300 articles (100 in peer-reviewed journals) and has been invited for more than 30 keynote speeches around the globe. He is involved in the boards of a number of international scientific organizations, holding positions including the following:

- President of the Commission Internationale de l'Organisation Scientifique du Travail en Agriculture, (Founded in Paris, 1950) (CIOSTA), for the periods 2011–2013 and 2017–2019;
- Chair of Section V (Systems Management) of the International Commission of Agricultural and Biosystems Engineering (CIGR), 2020–2022;
- Member of the National (Greece) Sectoral Scientific Council (SSC) for Energy;
- Chair of the European Federation for Information Technologies in Agriculture (EFITA), 2019–2021.

Charisios Achillas is Assistant Professor at the International Hellenic University, Department of Supply Chain Management. He graduated in 1999 from the Department of Engineering (Aristotle University of Thessaloniki), with a degree in Mechanical Engineering. He continued his studies with an MSc in Engineering Project Management in 2001 (UMIST, Manchester, UK). In 2009, he received his PhD Doctorate in the field of reverse logistics. Dr. Achillas is a senior researcher at the Institute of Bio-Economy and Agri-Technology, Center for Research & Technology Hellas (CERTH) and the Laboratory of Heat Transfer and Environmental Engineering, Department of Mechanical Engineering, Aristotle University of Thessaloniki. Since 2003, Dr. Achillas has been involved in science, research and development, from technical development to coordination, including project and financial management. His work has flourished in participating in more than 40 research projects (the majority being EU-funded), dealing mostly with environmental engineering, reverse logistics and sustainable development. Dr. Achillas is the author of more than 170 scientific publications. His research interests span the fields of environmental management, sustainable development and circular economy, with a primary focus on reverse logistics and agri-chains.

Preface to "Supply Chain Management for Bioenergy and Bioresources"

In the modern world, the competitiveness of bioenergy- and/or bioresources-related activities heavily depends on the effectiveness of supply chain management. A large number of multidisciplinary topics are involved in the bioresources and bioenergy production fields. Although the technical issues that are related with the topic are well-discussed and do not represent major barriers, supply chain management issues, such as design of the network, collection, storage or transportation of bioresources, are still considered as fundamental questions that need to be answered to enable the optimal exploitation of bioenergy and bioresources. Moreover, modeling of material and energy flows; identification of the dynamic character of the supply chains; available reverse logistics (waste management) alternatives; economic, social and environmental sustainability of bioresource supply chains; novelty in the applied business models; and decision support frameworks towards efficient supply chain management for bioenergy and bioresources present critical operational sustainability issues and business-making potential.

This Special Issue, entitled "Supply Chain Management for Bioenergy and Bioresources", includes one extensive review on yellow and woody biomass supply-chain management, together with six original papers which span around a number of innovative, multifaceted, technical developments that are related to all different echelons of supply chain management for bioenergy and bioresources. More specifically, in their work ("Green, Yellow, and Woody Biomass Supply-Chain Management: A Review"), Rodias et al. present a comprehensive review on research studies targeting biomass supply-chain management advancements related to three types of biomass sources, namely green biomass sources (such as perennial grasses), yellow biomass sources (such as crop residues) and woody biomass sources (such as willow). From their review, it becomes evident that the presented up-to-date trends on biomass supply-chain management and the potential for future advanced application approaches play a crucial role in business and sustainability efficiency of a biomass supply chain.

The Special Issue is also enriched by the following six studies: (1) Lampridi et al. ("Energy Footprint of Mechanized Agricultural Operations") present a modeling methodology for the precise calculation of the energy cost of performing an agricultural operation, with their model incorporating operational management into the calculation, while simultaneously considering the commercially available machinery (implements and tractors); (2) Vlachokostas et al. ("Decision Support System to Implement Units of Alternative Biowaste Treatment for Producing Bioenergy and Boosting Local Bioeconomy") propose a generic methodological scheme based on multicriteria analysis for the development of small-, medium- or large-scale units of alternative biowaste treatment, with an emphasis on the production of bioenergy and other bioproducts, taking into account environmental, economic and social criteria to support robust decision-making; (3) Papageorgiou et al. ("Forecasting of Day-Ahead Natural Gas Consumption Demand in Greece Using Adaptive Neuro-Fuzzy Inference System") examine the application of neuro-fuzzy models, so as to develop a real, accurate natural gas (NG) prediction model for Greece, thus providing a fast and efficient tool for utterly accurate predictions of future short-term natural gas demand; (4) Vamvanoulas et al. ("Dry Above Ground Biomass for a Soybean Crop Using an Empirical Model in Greece") propose a new empirical equation for the estimation of daily dry above ground biomass for a hybrid of soybean, which presents a useful tool for estimations without using destructive sampling; (5) Papageorgiou et al. ("Decision-Making

Process for Photovoltaic Solar Energy Sector Development using Fuzzy Cognitive Map Technique") focus on the investigation of certain factors and their influence on the development of Brazilian photovoltaic solar energy with the help of fuzzy cognitive maps, an established methodology for scenario analysis and management in diverse domains, inheriting the advancements of fuzzy logic and neural networks; (6) Wu et al. ("A Cloud-Based In-Field Fleet Coordination System for Multiple Operations") analyze the structure and composition of an auto-steering-based collaborative operating system for fleet management that is developed to realize an in-field flow-shop working mode often adopted by the scaled agricultural machinery cooperatives.

In conclusion, this Special Issue seeks to contribute to the bioenergy and bioresources agenda through enhanced scientific and multidisciplinary knowledge that may boost the performance efficiency of supply chain management and support the decision-making process of stakeholders. We are confident that the papers included in the present Special Issue provide interesting case studies that may motivate researchers and encourage additional efforts in supply chain management for bioenergy and bioresources.

Dionysis Bochtis, Charisios Achillas
Special Issue Editors

energies

MDPI

Review

Green, Yellow, and Woody Biomass Supply-Chain Management: A Review

Efthymios Rodias [1,2], Remigio Berruto [1,*], Dionysis Bochtis [2], Alessandro Sopegno [1] and Patrizia Busato [1]

1 Department of Agriculture, Forestry and Food Science (DISAFA), University of Turin, Largo Braccini 2, 10095 Grugliasco, Italy
2 Institute for Bio-economy and Agri-technology—IBO, Center for Research and Technology Hellas-CERTH, 10th km Thessalonikis-Thermis, BALKAN Center, BLDG D, 57001 Thermi, Greece
* Correspondence: remigio.berruto@unito.it; Tel.: +39-011-670-8596

Received: 8 July 2019; Accepted: 1 August 2019; Published: 6 August 2019

Abstract: Various sources of biomass contribute significantly in energy production globally given a series of constraints in its primary production. Green biomass sources (such as perennial grasses), yellow biomass sources (such as crop residues), and woody biomass sources (such as willow) represent the three pillars in biomass production by crops. In this paper, we conducted a comprehensive review on research studies targeted to advancements at biomass supply-chain management in connection to these three types of biomass sources. A framework that classifies the works in problem-based and methodology-based approaches was followed. Results show the use of modern technological means and tools in current management-related problems. From the review, it is evident that the presented up-to-date trends on biomass supply-chain management and the potential for future advanced approach applications play a crucial role on business and sustainability efficiency of biomass supply chain.

Keywords: supply-chain design; strategic planning; operational planning; energy crop production; crop residue

1. Introduction

For the generation of any product, a sequence of processes connected to design, decision making, and execution, and a series of financial, information, and material flows are performed throughout different stages of the production. The different stages of the production constitute an integrated system called supply chain. The basic aim for the successful design of a supply chain is to meet the requirements of the final customer regarding a specific product. The supply-chain concept for agri-food and biomass relates to not only manufacturing and retailing sectors but also to agricultural sector.

In recent decades, energy crops constitute a highly-potential share among crops taking into account the need for greener energy production. Energy crops are crops that are cultivated for biomass, biogas, or other biofuels (e.g., biodiesel, bioethanol) production. They are mostly green crops that come from wild nature, such as perennial grasses with high potential for bioenergy production. Green-type biomass includes crops such as *Miscanthus*, *Panicum virgatum* (also known as switchgrass), *Arundo donax*, etc. At the same level, yellow biomass refers to crop residues that come from any crop and represent another category of biomass production related to feedstock. Examples of yellow biomass are corn stover, wheat straw, etc. It should be mentioned that the main sources of biogas production are the energy crops and the use of agricultural residues [1]. On top of this, there is a variety of woody crops that contribute significantly to biomass energy production globally. Woody biomass is any biomass that is connected to wood sources. Examples of woody biomass sources are willow, poplar short-rotation coppice, etc. All of them have various and complex constraints regarding the entire management policy and practices that should be followed for the optimal biomass production.

The main challenges regarding supply chain issues on each biomass type category, as they are presented above, are different. Challenges related to green-type biomass include any grass-type crop operational issues, including particular issues in harvesting and handling (such as optimal scheduling), and less on other processes (such as soil cultivation or fertilization) due to their easy adaptability to various environments. On the other hand, yellow-type biomass requirements include optimal collection, handling and transportation processes. This incorporates on-time scheduling of collection and transportation and optimal task execution in cases where multiple fields are covered. Challenges on woody biomass sources are different from the other two types given its operational processes particularities. The woody energy crops require different crop establishment, cultivation, harvesting, and transportation processes. Of course, some operational issues would be similar with the other two types (such as scheduling of operations), but there are technical issues that are solved in different ways. A short example regarding collection and transportation would be about harvesting of green or yellow biomass in bale form compared with woody crops that whole trees are collected.

At this time, supply-chain management (SCM) in agricultural production, handling, and transportation processes is vital and there are always various issues that should be faced through better SCM. There is a large amount of research works regarding SCM of green, yellow, and woody biomasses. The work here, targets on an up-to-date literature review on recent publications on green, yellow, and woody biomass SCM.

The main challenges for the creation of this review were, firstly, to underline variation practically in different approaches for specific existing problems in SCM, secondly, to provoke the development of supply-chain management on the specific target group by proposing possible solutions on various upcoming matters and, finally, to provide a brief review of various followed practices/methodologies and their effects on the SCM.

There are previous reviews on the supply-chain management in agricultural processes regarding green, yellow, and woody biomass types. A biomass supply chain evaluation and optimization are suggested by a literature review regarding forest feedstock [2]. A systematic review was presented in order to present the key factors throughout the biomass supply chain of green and woody crops that affect the application of bioenergy buffers in complex bioenergy production systems [3]. A wider review about biofuel SCM is conducted under the objective of uncertainties and sustainability issues [4]. On the opposite side, a more practical review was presented regarding many types of mathematical models in bioenergy crops production, including both energy crops production processes and transportation but also biorefinery/biomass conversion modeling processes [5]. Even though the scientific contribution of these reviews is highly important, to the knowledge of the authors, there is no recent review regarding chain management aspects for green, yellow, and woody supply.

The objective of this paper is to highlight and focus on green, yellow, and woody biomass supply-chain management research works (52 studies), and, as a second step, to create a classification in order to propose opportunities for further research by focusing on research gaps and identified issues needed to be tackled.

2. Supply-Chain Management Definition

In order to conclude to a successful definition of SCM, i is important to make a short comparison between traditional management (TM) practices and SCM practices. Under the financial concept, by TM a reduction in a company's costs may be achieved, while by SCM a whole-chain cost efficacy will be obtained. Regarding data exchange and monitoring information, the first case is limited on the business's own needs, while in SCM it can be extended for whole-chain planning and/or monitoring processes. Another point is the coordination between different levels of a channel, where in TM there is only a single contact for the interchange among the channel pairs, while by SCM multiple contacts and coordination between various businesses and levels of channels can be accomplished. Finally, there is a number of risks and rewards that cannot be shared under the TM philosophy, but by SCM all the risks and rewards are shared in a long-term period. The above-mentioned comparison is only a small sample of the differences between TM and SCM, suggesting the need for assimilating more and more SCM practices.

There are various ways to describe and define SCM. For the purpose of this paper a definition of the term regarding crop production processes would be: Supply-chain management is the integrated planning of in-field and/or logistics operations, application of these operations, coordination between the different levels of the channel(s) and, finally, control of all processes and necessary activities in order to produce and transport, in the most efficient way, the products that finally will satisfy the requirements of a given market [6].

Given this definition, we could set that all the in-field and logistics processes are included in the term supply chain, not only as a physical operation but also as a decision-making activity, both associated by material flows and exchange data and, as a consequence, the correlated financial/energy flows. In this light, the supply chain includes not only the producer and its suppliers, but also includes the processing units, logistics operators, warehouses, etc.

3. Review Methodology

The methodology followed in this review includes a series of theoretical considerations taken into account in the pre-processing stage. Throughout our analysis, the included steps (as also presented in Figure 1) are the following:

- Step 1: Development of the review protocol in terms of the eligibility criteria considering a single published research article as the set analysis unit.
- Step 2: Search for research studies and select the ones satisfying the eligibility criteria.
- Step 3: Definition of the classification framework to be applied in the literature review in order to classify the material and build the structure. Four classes (green, yellow, woody, and multiple-type biomass) are applied here with two sub-classes (i.e., the problem-based class and the methodology/approach-based class).
- Step 4: Selection of studies to be included in each classification within the framework.
- Step 5: Analysis of the selected studies and creation of a short summary of each individual work allocated to the corresponding class.
- Step 6: Representation of results by studies comparison.

3.1. Eligibility Criteria

Studies related to green, yellow, woody, and multiple-type biomass SCM are included in this study. At the same time, as woody biomass is only referred to woody energy crops (such as willow), publications related to forest species or forest waste biomass are excluded from the scope of this review. The SCM-related tasks are included in the wide boundary in which this review's scope is subjected, as presented in Figure 2. For the scope of this study, the included publications should be related to SCM, and should be referred mainly to the crop production processes and/or the transportation from farm to the field and/or from field to the plant. Any further biomass conversion operations or processing was considered to be out of the scope of this review. However, publications that refer to individual operations and not to inter-connected parts of the supply chain were not taken into account.

A number of literature-related eligibility criteria are set. These criteria include: (1) The work should be published in English language, (2) the included studies should be research articles published in peer-review journals, and (3) they should have been published within the last five years (current one included; i.e., from 2015 up to present). Publications in journals that are not research articles, such as reviews or short communications are excluded from this review paper.

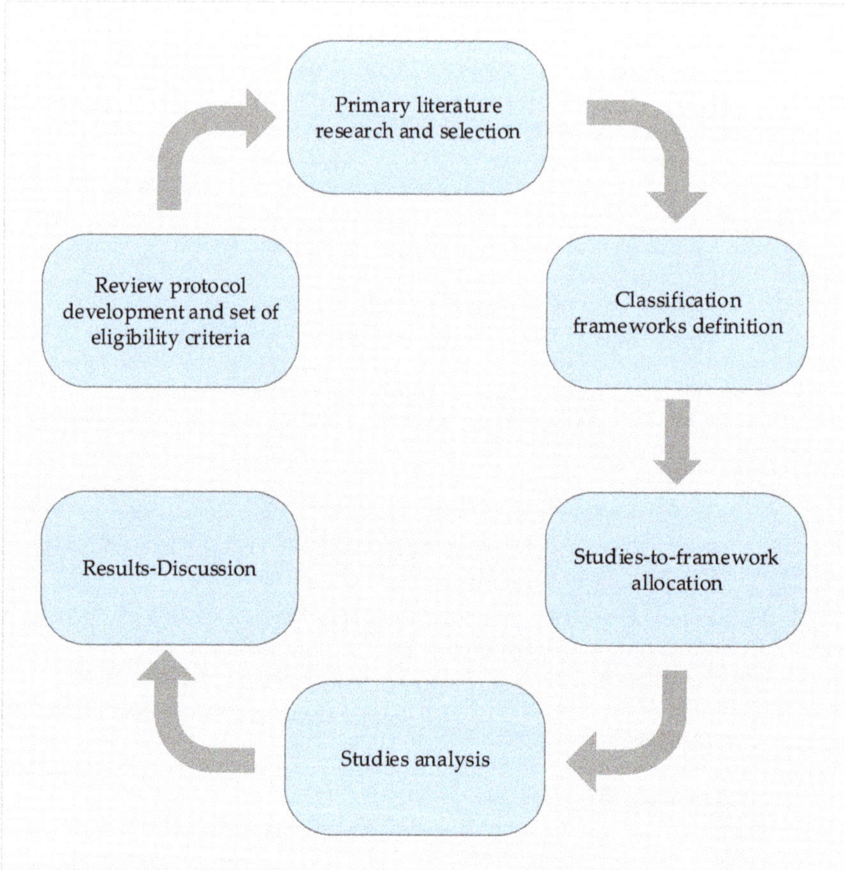

Figure 1. The summary of the review process.

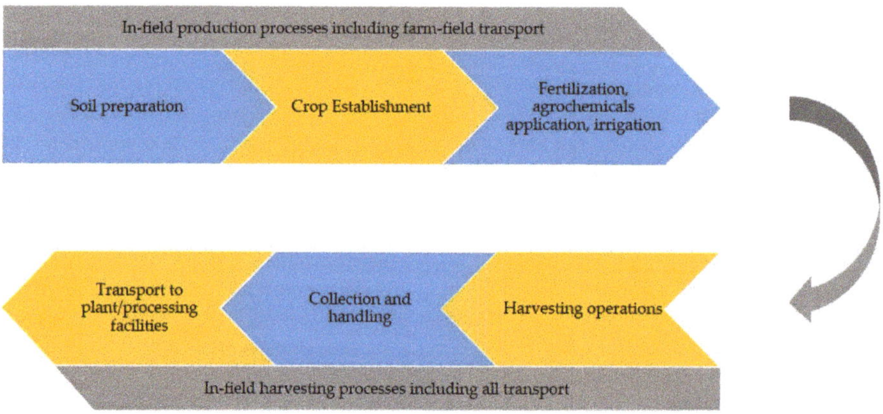

Figure 2. Processes included in this review's boundary.

3.2. Information Sources

There are a total of 52 studies included, which withdrawn from the electronic databases: Web Of Science, ScienceDirect, Wiley Online Library, and SpringerLink. The primary searches were implemented in the mid-end of January 2019. It is possible that any newly emerging publication after this date is not included. Nonetheless, the keywords, their combinations, and the searching strings are saved in order to make the whole process replicable for future use.

In the present review, no grey literature is included (i.e., any literature produced in electronic or print format that has not been controlled by any commercial publisher, for example, technical or research reports, doctoral dissertations, etc.).

4. Biomass-Type Classification Frameworks and Analysis

The existing literature regarding green, yellow, or woody biomass sources was classified in the following categories: (1) Literature focusing on supply-chain strategic planning, and (2) literature regarding operational planning for a series of operations (or the whole system) throughout the supply chain. In the term "supply-chain strategic planning", system productivity or life cycle analysis (LCA) (or energy consumption/balance assessment) of the biomass crop under study was considered. In a similar way, any financial/economics evaluation of a certain crop production/operational processes would be included under supply-chain strategic planning. However, these categories are not rigid and there is high possibility a particular work could be incorporated in both of them. A brief summary of the literature allocated to these categories according to the specific biomass type reported is presented below.

4.1. Green Biomass

4.1.1. Strategic Planning

Studies related to green biomass feedstock that include assessment of the system productivity or LCA analysis (including sub-categories such as energy consumption/balance, etc.) or financial evaluation are incorporated in this category. Apart from this, studies that include innovative design strategies throughout supply chain are also under consideration here.

An up-to-date geographical information system (GIS)-based approach by using a supply-chain simulation model is presented [7]. The objective of the presented approach was the determination of the optimal locations of beet crops that maximizes the profit of bioethanol production plants. Another study on allocation of energy crops to dispersed field locations is the one presented by [8], where a crop-to-field allocation tool targeting the maximum energy gain of the system was proposed. In that study, energy consumption is considered for all in-field operations and transportation in the routes between farm-field and field-plant. The optimization problem was modeled both as a linear and as binary programming. Another simulation approach focuses on spatial geographical allocation of *Panicum virgatum* crop fields, storage facilities, and biorefineries were presented [9]. The work was based on a two-phase modeling process under the scope of economical sustainability of the supply chain components. This two-phase simulation included the implementation of the agricultural land management alternative with numerical assessment criteria (ALMANAC) model for crop productivity simulation combined with the AnyLogic simulation model that is capable, among others, of finding the optimal locations for biomass storage facilities.

In [10], alternative supply chain configuration scenarios in *Miscanthus* harvesting and transport were compared, financially optimized, and assessed from a sustainability perspective for different annual demand in biomass, yield, and time of harvesting, among others. For the same energy crop, a computational tool was presented by [11], based on an in-depth analysis and energy requirements estimation of individual *Miscanthus* fields, including all the in-field operations and transportation. An economic approach based on the implementation of GIS was implemented to evaluate green biomass (*Miscanthus* and *Panicum virgatum*) SCM under different structures of the supply chain [12]. The scale of examination was downsized from county-level to field-level in order to more easily find the suitable individual areas for biomass production and, in this way, allowing microeconomic evaluation and modeling.

Miscanthus was also under the scope of study in terms of supply-chain strategic planning considering the production operations [13]. A simulation model called Simulateur mulTIdiscplinaire pour les Cultures Standard (STICS) was presented (as an improvement of an existing one) for the accurate prediction of the produced biomass in the long-term (up to 20 years from planting time) in various case studies. On top of this, the authors evaluated the nitrogen content in different cropping environments and in-field management practices. An approach based on soil maps, climatic data, and observed yields of *Miscanthus* fields was presented by [14] and in combination with GIS data concluded in the perspective of a decision support tool development for optimal supply-chain strategic planning. An optimization tool was also evaluated in the *Miscanthus* production supply chain and presented promising results as a decision support tool for farmers—in crop cultivation strategy development—and for policy makers—in monitoring and improvement of supporting practices [14].

A combined financial and energy requirements analysis was conducted by [15] for harvesting and transportation operations in *Arundo donax* production systems. Targeting the optimal strategic planning of supply chain, different operational alternatives were proposed.

Sugarcane is evaluated in [16] study, which focuses on sugarcane SCM and also on green harvesting residues management under different operational practices. The authors developed a simulation model to present the biomass flow through the whole supply chain (in-field and transport operations). Except for operational constraints, they also took into account weather and geographical constraints to identify possible bottlenecks in the supply chain and biomass availability, such as non-synchronized harvesting, handling, and transporting operations.

Triticale is a hybrid of wheat and rye not widely considered as a crop for bioenergy production. A study that focused on the assessment of triticale as a potential biorefinery feedstock is the one presented by [17]. They introduced improved harvesting methods that have a positive effect on costs reduction and availability secure of high-quality biomass in the long-run. A number of autochthonous perennial grasses, as potential crops for biomass production, were evaluated in terms of supply chain efficiency [18]. Authors conducted a four-year experiment to examine parameters such as energy efficiency, crop yield, and water-use efficiency.

Energy balance estimation comes to the center of interest for complex biomass production systems, as presented by [19]. More analytically, they presented an assessment process for the energy balance of multiple-crops and multiple-fields systems by implementing a web-based tool. Three different crops were evaluated as a case study, namely corn silo, wheat, and rapeseed. In [20], multiple-production systems were also evaluated by developing and implementing a comparative computational tool potentially applied to any set of given crops, given energy-related or production-related input, and according to any specific production practices. A study that refers to different bioenergy cropping systems is presented by [21]. They evaluated the productivity of selected systems by using experimental data that were further analyzed by a simulation tool to identify potential limitations of resource-use efficiency on biomass dry matter yield.

Carbon footprint assessment throughout the supply chain is included also in supply-chain strategic planning. An integrated assessment of commercial crops (such as *Maize*) with bioenergy crop production (such as switchgrass) can be an innovative way for biomass supply-chain strategic planning, as presented by [22]. Authors took into consideration the effect of SCM practices on cost, yield production, and system sustainability. A multiple-crop CO_2 annual emissions estimation regarding cultivation processes was conducted by [23]. In addition to this, they made a long-term (for a 30-year period) prediction of the CO_2 emissions for the selected crops in order to identify the most sustainable crop(s), from the environmental point of view, for bioethanol production. Environmental impact assessment of multiple perennial green biomass crops in marginal land was presented by [24], based on crop comparison under specific conditions and further development and improvement of the supply chain.

4.1.2. Operational Planning

A multiple-optimization strategy was followed considering the impacts of operational management on minimization of costs, maximum, yields, and sustainability of *Panicum virgatum* supply chain [22]. For this scope, authors modeled the integration of *Panicum virgatum* on a real corn production field and its effects on profitability, productivity, and environmental improvements of the system by using mainly the landscape environmental assessment framework (LEAF) tool (Version 2.0, United States).

Arundo donax is also the target energy crop in study [25], where authors provided an agronomical assessment testing on the crop structure and regrowth potential as they are affected by the harvesting time. They further compared different harvesting alternatives and evaluated the biomass quality based on the harvesting parameters, such as harvesting time and harvesting frequency.

Sweet sorghum and sugar beet are two important energy crops for bioethanol production. In this light, many studies have focused on the SCM of these crops. The biomass yield of sweet sorghum and sugar beet was estimated in an experimental study in southern Italy [26]. In the same work, the energy performance under different in-field operational management scenarios was evaluated for these two crops. The effect of three levels of shade (low, medium, and high) on *Maize* production (such as growth and yield) that is cultivated for biogas production was evaluated in [27]. These levels of artificial shading were presented to affect the biogas-related parameters (such as leaf area index and energy availability for plant growth) and the final biogas and methane yield. In a *Miscanthus* production system, biomass production, costs, and supply-chain constraints are considered in a whole supply chain financial optimization strategy [10].

There is also reference considering specific aspects of supply-chain operational issues in green biomass crops. The optimized design of the in-field operations following optimal route planning (B-patterns) under the objective of minimizing energy cost was assessed in two green cropping biomass production systems (*Miscanthus* and *Panicum virgatum*) [28].

4.2. Yellow Biomass

4.2.1. Strategic Planning

The establishment of biogas plants in the optimal location is directly connected to crop residue-related parameters (such as quantity, accessibility, weather conditions, etc.). Regarding the strategic planning of yellow biomass supply chain, the models multiple linear regression and artificial neural networks (ANNs) were implemented for the estimation of available crop residues (specifically, corn stover and wheat straw) in multiple sites [29]. In the same work, potential suitable locations for bioenergy plants are identified by this approach, and subsequently, the optimal plant location is suggested together with the biomass delivered cost. For the same crop residues, authors in [30] calculated the potential sustainable yellow biomass quantities while maintaining soil productivity and health. They focused on large-scale bioenergy applications and proposed the establishment of sustainable removal rates of residues and supply chain cost for various regions.

A simulation-based model that considers multiple-locations assessment and selects the optimal location for bioethanol plants based on parameters such as wheat biomass density, supply chain network, etc., was presented by [31]. The focus was on the financial analysis and environmental impact assessment.

Regarding crop residues from corn production (stover), different biomass handling and transportation scenarios were evaluated under an LCA analysis in [32]. The selected scenarios included combinations of biomass handling (in bale form or pelletized form) and either storage in an intermediate depot or by directly transportation from field to the biorefinery.

Cotton stalks represent another type of crop residue biomass. In [33] an integrated GIS and an ANN high spatial resolution model was developed to assess available cotton stalks harvesting and transportation. In addition, by GIS analysis the suitable biorefinery locations were selected under the criterion of the minimum total transport distance and the delivered cost. Finally, the estimation of spatial and temporal variations of potential cotton stalks in the United States was presented by [34], including also an assessment of different SCM practices for cotton stalks under an engineering economics approach.

A GIS-based model taking into account various features connected to a geographic region in order to determine the location of biomass-processing facilities considering also uncertainty factors [35]. These factors are connected to the availability of crop residues, other environmental, financial, weather, or even social constraints. In this study, residues coming from various crops are used, such as corn, sorghum, sugarcane, wheat, barley, agave, rice, and pecan nut. The development of a spatial decision tool is also the main objective in [36], with the aim of advancing this research field. The developed tool is useful at the strategy development level where the decision making is conducted. The optimal alternative solutions or combination of them were determined by identifying the best regions for the establishment of biomass plants, according to the crop residue availability, by including also factors such as geographical features, accessibility of the crop residues points, and the complexity and structure of the logistics network. Multiple types of crop residue biomass were also under consideration in [37]. Authors presented a location-route collaborative optimization model that they used to design the multiple residues collection network and corresponding routes. The yellow biomass considered was stalk, leaves, etc., from crops such as cereals and beans, tobacco stalks and leaves, stalk from beet, sugarcane, among others. The sites were depicted by nodes and thus the node capacity of the network and the logistics cost were the basis for the main modeling structure. On top of that, they proposed and built a mathematical algorithm to solve the location-routing problem under specific constraints.

4.2.2. Operational Planning

Fruit orchards, vineyards, and olive trees are of high biomass interest given the produced residues in the field. On this, [38] presented an integrated framework for the development of a decision support tool for logistics operations of fruit trees, vineyards, and olive tree pruning and branches. In this work, the necessity of a "smart logistics system" is presented. Additionally, the components of this system under certain technical and other requirements and constraints were defined together with the determination of the information to be managed by the system. The olive tree-pruning biomass supply chain was accurately estimated and optimized on supply-chain development and modeling frameworks [39].

Innovative techniques that include quantification of pruning biomass residues coming from individual olive trees and based on LiDAR (light detection and ranging) data processing methods were used by [39]. Their work is of high value for supply-chain operational planning given that they focused on the accurate estimation of the biomass volume and distribution and other aspects such as supply-chain modeling and development.

4.3. *Woody Biomass*

4.3.1. Strategic Planning

SCM is quite significant also for woody biomass considering strategic planning. Authors in [40] analyzed the viability of eucalyptus supply chain and further evaluated the optimal location of a power plant to be fed by wood. Their analysis was based on GIS databases to calculate available biomass under certain restrictions. Even though this work is one among many that target the optimal bioenergy plant localization, their methodology includes calculation of optimized costs and CO_2 emissions throughout the supply chain considering local and regional data. Authors in [41] developed a discrete event simulation model for techno-economic assessment of harvesting, handling, and transport operations for the supply of short-rotation coppice willow (SRCW) to energy conversion plants. Their main objective was to suggest cost-effective supply chain solutions in order to deliver all-year-round SRCW to the biomass plant.

Poplar short-rotation coppice (SRC) belongs also to woody species under consideration for biomass production. In the work presented in [42], the most important environmental impacts of wood chips production from poplar SRC on marginal land were provided in detail, and a comparison was made between common SCM practices and alternative SCM practices in terms of modified harvesting and handling systems, irrigation, and fertilization.

Eucalyptus is included in the woody biomass SCM category and is considered one of the most commercial woody crops cultivated for bioenergy production. In [43], the economic viability of various operational practices regarding whole eucalyptus tree trunk production was analyzed under varying production and economical parameters. To analyze risk management a Monte Carlo analysis was run. Stochastic models also included different planting densities and arrangements in order to provide a framework for assessment of financial risk applied in decision analysis for bioenergy crop investments.

Willow was under consideration in study [44], where part of the supply chain was modeled and evaluated (harvesting, collecting, and transportation). Authors run the integrated biomass supply analysis and logistics (IBSAL) simulation model (Oak Ridge National Laboratory, Oak Ridge, TN, United States) in a significant number of SR willow fields and evaluated the effect of the major input factors (i.e., parcel size, field shape, crop yield, field-to-plant distance, and type of collection machinery) in the performance of the system.

An innovative mobile application for application in mostly woody biomass supply-chain strategic planning was suggested by [45]. The objective was to create a useful tool for the farmers in order to be able to evaluate the energy potential of tree pruning in their own orchards, share the information regarding the specific region that the orchard belongs, and match this data with the availability for heating energy in agro-industrial or other buildings.

4.3.2. Operational Planning

In [40], optimization of supply chain was pointed out, targeting the optimal location of wood-fired plants under specific constraints. SRC willow, as it has been presented above in [41], is under the scope of supply-chain operational planning (regarding harvesting and handling operations) in the supply chain from field to plant. Additionally, the impact of the most significant input factors was evaluated regarding the total willow supply chain performance [44].

Poplar tree production was also evaluated on part of the supply chain related to cultivation and harvesting operations [46]. The main objective was to point out the environmental impact of poplar trees cultivated in various soil nutrient levels and suggest the optimal strategy. Similarly, in poplar SRC, optimization of cultivation and transportation practices through a financial and environmental analysis was presented by [47].

Regarding woodchip transportation on a short supply chain (up to 70 km distance), an analysis of the energy input and the carbon dioxide emissions was conducted by using two types of transportation means (i.e., agricultural tractors or industrial trucks) and carrying out a thorough analysis under parameters related to road design (such as, traffic lights, intersections, etc.), traffic conditions, and road surface conditions (such as rain, fog, ice) [48].

4.4. Multiple Biomass

4.4.1. Strategic Planning

Several studies include combination of green and yellow biomass supply-chain strategical and operational planning. A work that combines green and yellow biomass SCM based on agro-ecosystem models was presented by [49]. The main objective of the work was the supply chain assessment and design (including crop production and transportation) under the demand of a bioethanol plant with feedstock from three types of crops grown for different purposes, namely: crop residues; annual crops; and perennial crops (i.e., wheat straw, triticale, sorghum, and *Miscanthus*). The developed model focused on productivity measures and the environmental impact of the supply chain. Among others, the scope of the model was the minimization of the balance between food and feed crops. Under the scope of farmers' profit maximization and of environmental sustainability, a landscape approach for the assessment of multiple various-sized subfields was carried out in [50]. They evaluated opportunities coming from yellow biomass collection and transportation together with green biomass potential development on low-perspective fields that normally make no profit.

Supply-chain strategic planning is underlined by energy balance estimation of in-field and transportation operations research studies regarding green and woody biomass types, such as the ones presented by [19] and [51].

A comprehensive model that can be applied for strategic planning of green, yellow, and woody biomass supply chain was presented in [52], where authors developed a mixed-integer linear programming model that targeted the minimization of the establishment cost of the integrated supply chain under specific constraints (e.g., biomass availability, capacity of technology, etc.). Authors divided the supply chain in two parts: The biomass supply chain that referred to the part from field to biorefinery, and the liquid fuel supply chain that referred to flows of ethanol and blended fuels from biorefinery to gas station. The focus of the developed model was on supply-chain-related solutions and biomass-related solutions that may contribute to a more financially viable supply chain.

Finally, a CyberGIS platform (CyberGIS Gateway, Illinois, United States) was developed by [53] as a decision support tool for biomass supply-chain management and analysis. More specifically, authors described, in their model, how to integrate optimization within the CyberGIS platform in order to solve complex spatial decision-making problems throughout the supply chain.

4.4.2. Operational Planning

Another study combined green and yellow biomass assessment [54]. Authors evaluated both production and transportation operations of corn (stover) and *Panicum virgatum* under two different baling systems, and, in a second step, they compared the corresponding supply chains financially, in terms of energy cost, and in terms of GHG emissions.

Operational management practices were assessed within three different designed supply chains of green biomass in terms of operational times and costs [55]. In particular, three different approaches where simulated and the effect of yield uncertainty and machinery productivity were evaluated based on a sensitivity analysis approach. Authors in [56] proposed a simulation approach for the optimal biomass bales location. Various bale aggregation methods were simulated and evaluated in terms of the required in-field transport operations for different methods. Similarly, on [57] authors simulated the in-field layout of biomass bales, providing the desired bale stacks layout under various different parameters (such as field area and shape, yield, bale mass, etc.) for increasing the efficiency of biomass handling and transportation operations.

Scheduling of handling and transport operations is considered crucial in any biomass supply-chain strategic planning. In this light, a tool was developed for scheduling of field machinery used in these operations in geographically-dispersed fields under specific constraints [58]. In that work, authors presented an algorithm that creates individual working schedules for multiple machines that carry out more than on consecutive operations at multiple fields, taking into account the given readiness in each specific field on the specific timing. They used a basic Tabu Search method and a modified one that produces optimized work planning.

5. Methodological Approach-Based Classification Framework

The research studies, which were already classified above in two levels, can be further classified according to the methodological approach followed in each cited work. In this light, the four categories of included studies (i.e., green, yellow, woody, and multiple-type biomass) were kept the same as in the previous classification framework. The studies related to green-type biomass are summarized and presented in Table 1.

Table 1. Green-type biomass cited studies classified by crop and approach.

Cited Work	Crop(s)	Methods/Models	Criterion	Data Sets	Time Frame
[7]	Energy beet	Simulation	Maximization of profit of potential ethanol plants	Yields, production, and transportation costs, agricultural opportunity costs, ethanol production, and ethanol prices	N/A
[27]	*Maize*	Experimental approach evaluated by statistical analysis	Evaluation of biomass and biogas yields	Plant growth, yields, biogas, and methane yields	3 years
[17]	Triticale	Techno-economic computational model	Reduction of feedstock procurement operational costs, secured feedstock availability, and increased high-quality biomass supply	Triticale production data and biomass yields, commercial machinery data	1 year
[9]	*Panicum virgatum*	Two-phase simulation location-allocation modeling approach	Minimization of operational cost of the supply chain	Field, weather, and soil data sets, cost, demand, and price data and actual transport-related data	1 month
[18]	Wild perennial crops	Experimental approach together with statistical analysis	Biomass yields, biomass quality, and water-use efficiency	Meteorological data, soil water data, crop-related data	4 years
[10]	*Miscanthus*	Four-step assessment framework (i.e., design, optimize, assess, and compare biomass supply chains)	Economic optimization	Field and crop data, operational data, financial and energy-related data	1 year
[23]	Sugarcane, sugar beet, corn, rice and cassava	Experimental-based approach together with carbon flux assessment model	CO_2 emissions	Crop-related data, soil data, carbon emissions data, etc.	1 year
[12]	*Miscanthus* and *Panicum virgatum*	An economic, biophysical, and GIS modelling approach	Supply-chain structure and prices	Yields, climate data, field-related data, etc.	15 years
[11]	*Miscanthus*	Simulation	Energy requirements assessment	Operational data, machinery data, energy input, crop data, etc.	10 years
[15]	*Arundo donax*	A computational approach on economics and energy consumption based mainly on experimental data	Cost-effectiveness and environmental sustainability	Financial data, operating data, machinery and transport data, etc.	Annual
[14]	*Miscanthus*	A multiple regression modeling approach together with remote-sensing modeling approaches and experimental data	On-farm productivity	Soil water, climate data, georeferenced data, crop growth data, field and crop data, etc.	≥ 5 years

Table 1. *Cont.*

Cited Work	Crop(s)	Methods/Models	Criterion	Data Sets	Time Frame
[26]	Sugar beet and sweet sorghum	Simulation and experimental-based approach	Energy performance	Field and crop data, meteorological data, machinery data, operational data, etc.	3 years
[25]	*Arundo donax*	Experimental approach and statistical linear model	Biomass yield and quality	Climate data, productivity, and biometrical data, soil, site and crop data, etc.	2 years
[22]	*Panicum virgatum*	Multi-criteria decision analysis technique	Reduction of economic losses and maximization of environmental performance	Soil and crop data, profitability data, climate data, operational data, etc.	1 year
[8]	*Miscanthus*, *Panicum virgatum*, and *Arundo donax*	Binary and linear programming simulation models	Optimal energy performance	Crop and fields data, machinery data, material data, energy coefficients, etc.	10 years
[13]	*Miscanthus*	Simulation	Yield and N content	Crop and field features, soil, climate, agricultural practices, etc.	4–20 years
[16]	Sugarcane	Simulation	Cost, energy, and emissions	GIS data, weather data, farm data, operational data, etc.	Annual
[19]	Corn silo, wheat, and rapeseed	Web-based simulation tool	Energy balance	Fields and crop data, machinery and operational data, production means, etc.	N/A
[24]	*Miscanthus*, *Panicum virgatum*, *Arundo donax*, and cardoon	Simulation	Environmental impact	crop data, operational, production means, weather data, etc.	15 years
[20]	*Miscanthus*, *Panicum virgatum* and *Arundo donax*	Computational simulation tool	Energy cost	Fields and crops data, machinery and operational data, transportation, production means, and materials, etc.	10 years
[28]	*Miscanthus* and *Panicum virgatum*	Simulation	Energy consumption savings	Field and crop data, machinery data, operational data (turning radii, operating width), energy input, etc.	Annual

Another significant share of the potential biomass sources for bioenergy production regards yellow-type biomass. The studies that include agricultural residues SCM—only those that come from crop production and not any agricultural residue—are presented in brief in Table 2. These studies include various methods such as simulation and experimental methods, but also real-time artificial intelligence methods that are connected to machine learning technologies; technologies that have lately emerged in the agricultural sector [59].

Similarly, with the previous two literature categories, Table 3 regards studies correlated to woody biomass SCM approaches that come from woody crops but not common forest species. Moreover, in Table 4 the studies that refer to SCM of more than one different type of biomass feedstock are listed.

6. Discussion and Conclusions

In this review, 52 research studies were included in total, published in 24 different journals (Figure 3). "Biomass and Bioenergy" represent the most referenced journal in the current review followed by "BioEnergy Research" and "Journal of Cleaner Production".

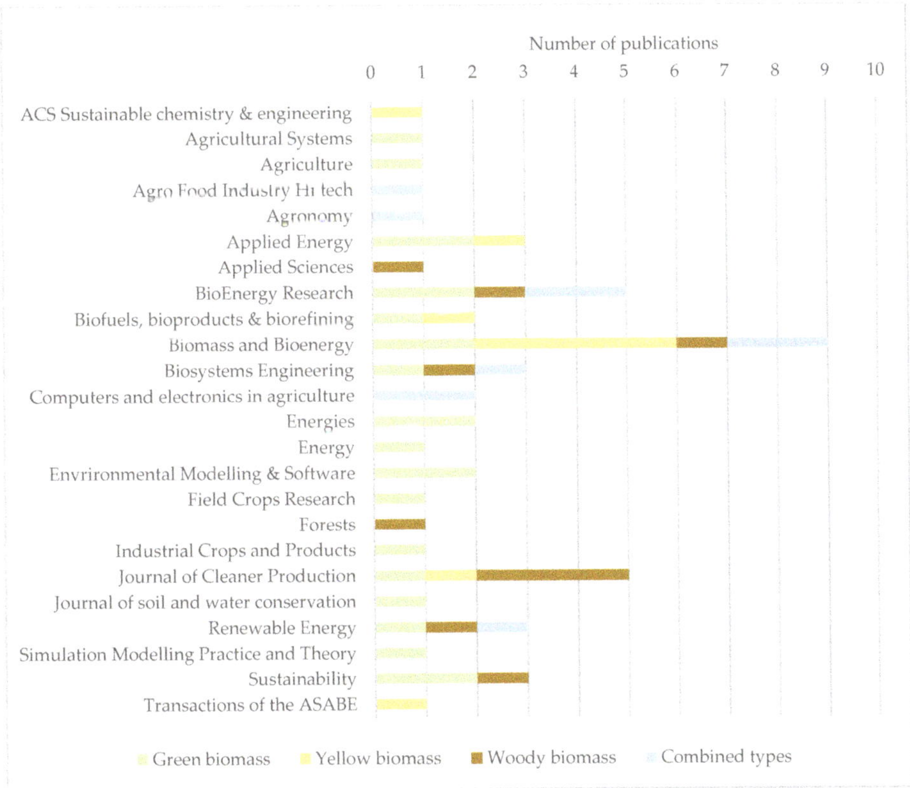

Figure 3. Number of papers extracted by each journal.

Table 2. Yellow-type biomass cited studies classified by crop residue and approach.

Cited Work	Type of Crop Residue(s)	Methods/Models	Criterion	Data Sets	Time Frame
[29]	Corn stover and wheat straw	Multiple linear regression and artificial neural networks (ANN) model	Crop residues availability, identification of optimal plant locations, and cost minimization	Crop yield, soil-related data, operational and financial data	Annual
[32]	Corn stover	Open LCA software-based modeling approach and Monte Carlo simulation	Environmental impact of supply chain versus densification and pelletization	Field trials and crop data, operational data, emission, and biomass related data	Annual
[31]	Wheat straw	Simulation-based approach	Minimum ethanol plant capital and operating costs in parallel with maximum profitability	Financial and environmental data, operating data	20 years
[33]	Cotton stalks	An integrated GIS and ANN prediction modeling approach and linear programming models	Yellow biomass availability, optimal plant location with the minimum supply-chain cost	Crop data, transport network data, weather data, geospatial data	5 years
[30]	Corn stover and wheat straw	Analytical modeling approach and simulation	Minimization of costs and environmental impact and maximization of energy efficiency	residual potential, machinery data, transport data, fields data, etc.	N/A
[34]	Cotton stalks	Analytical approach and simulation	Minimization of total delivered cost and energy input	Crop and field data, machinery data, operational data, etc.	Annual
[35]	Corn stover, sugarcane bagasse, sorghum straw, agave residue, wheat straw, rice straw, barley residue, and pecan nut shell	Geospatial optimization modeling approach	Optimal biomass processing location under uncertainty parameters	Spatial data, residual data, climatic data, terrain, and crop data	5 years
[36]	Cocoa crop residues	An integrated GIS-based fuzzy analytic hierarchy process (AHP) programming approach	Biomass availability, transportation, and slope feasibility	Crop yields, geospatial and land data, road data, accessibility, natural hazard, etc.	Annual
[39]	Olive trees pruning residues	Experimental approach based on LiDAR (light detection and ranging) technique	Quantification of olive trees pruning	Dendrometric data, field data, LiDAR data, etc.	N/A

Table 3. Woody-type biomass cited studies classified by crop and approach.

Cited Work	Crop(s)	Methods/Models	Criterion	Data Sets	Time Frame
[46]	Poplar	Simulation and experimental-based approach	GHG emissions	Field and crop data, operational and machinery data, emissions inputs, etc.	20 years
[47]	Poplar SRC	Modeling approach by using SimaPro tool and by using experimental data	Environmental and energy performance, economic viability	Yield and crop data, operational data, machinery data, energy and emissions input, economics, etc.	12 years
[43]	Eucalyptus	Financial analysis of different experimental silvicultural practices evaluated by Monte Carlo simulation	Investment analysis criteria such as net present value, internal rate of return, and profitability	Experimental, various economical and statistics-related data	3 years
[44]	Willow	IBSAL simulation model	Highest performance of the integrated system based on parcel size, field shape, crop yield, storage location, and collection equipment	Field trials data and commercial machinery data	5 years
[48]	N/A	Experimental-based approach combined with linear programming modeling	Energy requirements and CO_2 emissions	Vehicles-machinery data, road/traffic data, operational data	2 months
[40]	Eucalyptus	A computational approach based on WISDOM database and by using the network analyst tool	Minimization of costs and GHG emissions	Machine productivity data, operational data, yields, emissions data, road network data,	Annual
[41]	SRC Willow	Discrete event simulation modeling	Cost effectiveness	Weather data, transport data, yields, geographical conditions, soil water content, storage-related data, machinery data	Annual
[42]	Poplar SRC	Experimental approach based on growth model MoBiLE-PSIM and Umberto Software	Environmental impacts	Operating data, machinery data, field and crop data, growth data, etc.	20 years
[45]	Fruit trees	Mobile application development and performance approach	Biomass availability matching with heating energy requirements of agro-industrial buildings	Yields, energy requirements, climatic data, etc.	N/A
[38]	Fruit tree, vineyards, and olive grove prunings and branches from up-rooted trees	Smart Logistics System Prototype development and performance	Optimization of supply-chain performance	Biomass-related data, spatial data, weather data, etc.	N/A

Table 4. Multiple-type biomass cited studies classified by crop and approach.

Cited Work	Crop(s)/Residues	Methods/Models	Criterion	Data Sets	Time Frame
[37]	Woody and agricultural residues	Simulation and genetic algorithm development	Cost of transportation routes	Spatial data, yields, transportation and operating data, storage, etc.	N/A
[56]	N/A	Simulation and regression models	Logistics efficiency	Spatial data, crop and field data, transportation, etc.	N/A
[57]	N/A	Simulation	Logistics distances	Field and crop data, operational data, spatial data, transportation, etc.	N/A
[58]	N/A	Scheduling algorithm development	Field readiness	Crop and field data, operational data, transportation, etc.	N/A
[55]	N/A	Simulation	Task times and cost performance	Field data, transportation, operational data, machinery data, economics, etc.	N/A
[51]	Potential for up to 30 listed biomass crops	Web tool development	Yield and cost	Production data, fields and crops data, economics data, machinery, and operational data, etc.	N/A
[21]	Combination of various crops and crop residues	Experimental and simulation modeling	Productivity and resource-use efficiency	Data related to crop growth, field, weather and soil water	2 years
[52]	Various crops and crop residues	Mixed-integer linear programming optimization model	Minimization of the entire cost of the integrated bioethanol supply chain	Regional statistical data, crop-related data, operational cost data	N/A
[49]	Triticale, sorghum, and *Miscanthus*	Simulation-based approach	Optimal feedstock supply-chain strategic planning based on agro-ecosystem modeling	Data related to weather, crop management, soil, emissions, operations, etc.	20 years
[50]	Residues and energy crops	Simulation-based approach	Farmer's profitability and environmental sustainability	Crop rotations data, field-related data, soil data, weather data, operational and financial data	5 years
[53]	N/A	CyberGIS-enabled decision support platform development	Supply chain optimization	Spatial, agricultural (yields, production costs), and engineering/technology related data (such as transportation and operating data, etc.)	N/A
[54]	Corn stover and *Panicum virgatum*	Simulation	Minimization of cost and energy consumption	Machinery data, operational data, financial and energy data, yields, etc.	10 years

Generally, the use of final biomass product may vary, either for heating or electricity production closer to its primary natural form (i.e., pure biomass), either by converting biomass to biofuels (such as bioethanol, biodiesel, or biogas) through biorefining processes. Here, three main categories (namely biomass, biogas, bioethanol) as types of final product were pointed out in the included literature and presented in Figure 4, classified further to each different type of biomass. According to the literature, in the case of multiple types of biomass sources, the final product mostly leads to direct biomass use and less to bioethanol production. In parallel, literature shows that green and yellow biomass types may be equally used for biogas production. Regarding the bioethanol-related works, all types of biomass are referenced except woody type. This is also representative of the existing situation in bioethanol and biogas production industries.

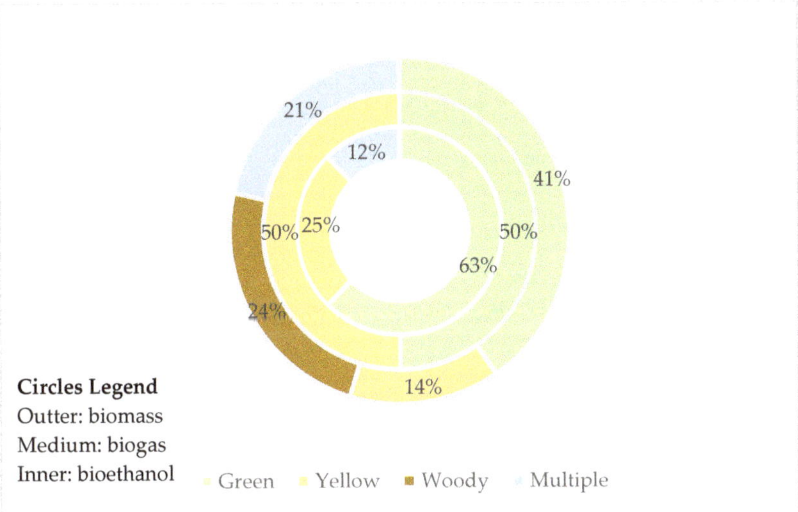

Figure 4. Biomass type in relation to the final product.

The reviewed literature was assessed under the methodological approach defined above and the results were presented in Tables 1–4 for each biomass-type. An aggregated representation of this categorization is presented in Figure 5. In general, various types of simulation approaches were used in almost 43% of the total studies of this review, while 26% used experimental methodological approaches and 31% used other types of approaches, such as online tools, web applications, etc.

As previously mentioned, the parts of the supply chain that were included in this review correspond to the stages from in-field production processes up to transportation to the biomass processing facilities, including any intermediate steps of the chain. However, the steps of the supply chain included in each conducted study varied. On this, eleven different combinations of supply chain steps were found in total in all biomass types. The number of works allocated to each one of these categories is presented in Figure 6. There is a significant amount of studies—19% of total—that correspond to in-field production and biomass transport stages, while 17% of the studies are targeted to in-field processes.

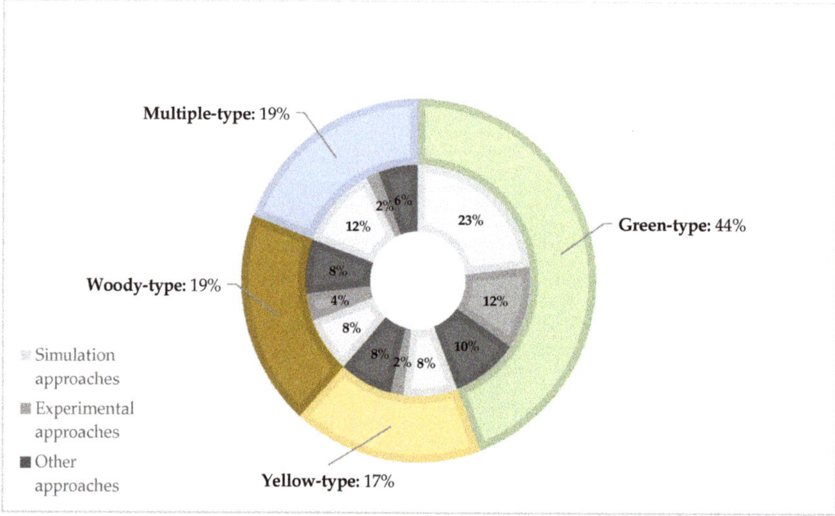

Figure 5. Biomass type and methodological approaches correlation.

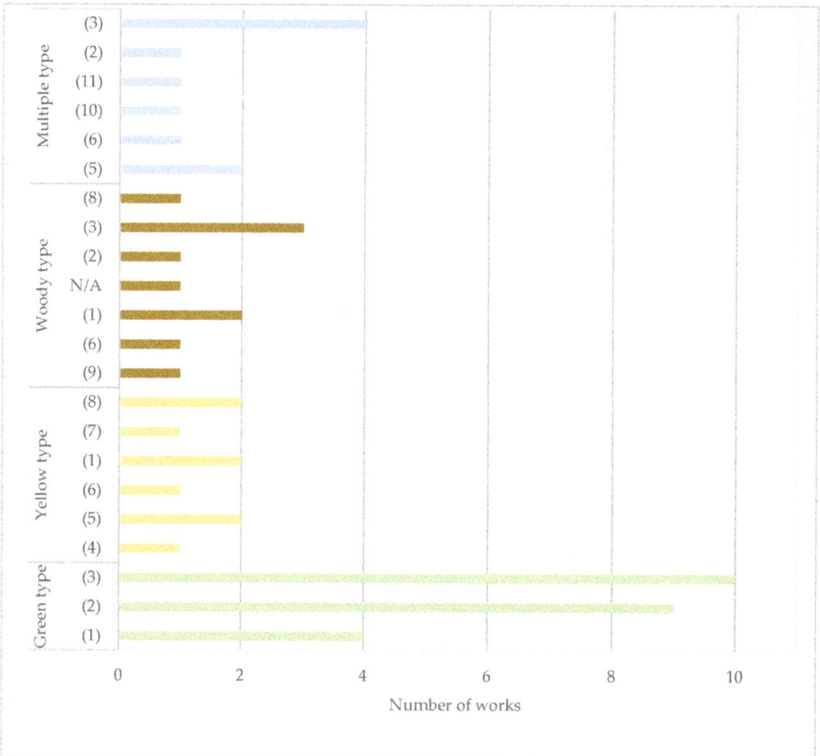

Figure 6. Number of works allocated to various parts of supply chain: **(1)** Harvest and transport; **(2)** production; **(3)** production and transport; **(4)** collection and quantification; **(5)** collection and transport; **(6)** handling and transport; **(7)** harvesting, collection, transport, and handling; **(8)** transport; **(9)** harvesting, collection, and transportation; **(10)** harvesting and handling; **(11)** harvesting, handling, and transport.

Among the articles, three supply chain categories are related to energy crops biomass production and six categories are connected to crop residues biomass supply-chain management. On top of this, there are seven categories related to woody energy crops biomass supply chain and six categories are presented to connect with studies that take into account more than one type of biomass (such as combined green- and yellow-type).

Overall, 52% of the referenced publications are related to strategic planning of supply chain, 29% of the works are related to operational planning, while 19% of the works present combined methods (targeting strategic and operational planning of supply chain). Green biomass-related literature includes approaches related to strategic planning (12 works), operational planning (five publications), while there are four works that refer to both strategic and operational planning tasks of supply chain. Regarding yellow biomass literature, five and one approaches are, respectively, targeted to strategic and operational planning, while three works include both strategical and operational planning Similarly, for woody biomass literature, three works are related to strategic planning, four are related to operational planning, and three are related to approaches that combine strategic and operational planning of biomass supply chain. Finally, in literature that includes multiple biomass types, strategic planning of supply chain is targeted seven times, five works refer to operational planning, and three works are related to a combination of strategic and operational planning approaches.

Regarding the methodology-based framework, as it was presented in Figure 5, simulation techniques are the mostly used methodological tool in green-type biomass and multiple-type biomass systems. In yellow and woody type biomass, experimental approaches and other methods (such as analytical models, etc.) are the main methods.

Combining simulation methodological approaches to experiment-based methods in any kind of biomass type may result in the real evolution of these systems under the scope of optimal SCM. Moreover, in the future, methodological tools and other types of approaches are expected to contribute to this evolution by developing innovative real-time tools that would be useful, especially in solving problems in complex biomass systems. There is already a trend in this direction but this is quite dynamic and can be widely expanded more in many levels, both for the needs of the mentioned crops and systems but also in other biomass systems. By this review, an integration of different approaches is suggested for further evolution in currently-used methodology under the scope of production levels increase, sustainability enhancement of the systems at hand, and for ensuring bio-based product quality.

Author Contributions: Writing—Original Draft Preparation, E.R., D.B. and R.B.; Methodology, E.R., D.B. and P.B.; Investigation, E.R., D.B. and A.S.; Conceptualization D.B. and R.B.; Writing—Review and Editing, E.R. and D.B.; Supervision, R.B. and D.B.

Funding: This research received no external funding.

Conflicts of Interest: The authors declare no conflicts of interest.

References

1. Hossain, G.S.; Liu, L.; Du, G.C. Industrial Bioprocesses and the Biorefinery Concept. In *Current Developments in Biotechnology and Bioengineering: Bioprocesses, Bioreactors and Controls*; Elsevier: Amsterdam, The Netherlands, 2016; pp. 3–27. ISBN 9780444636744.
2. Cambero, C.; Sowlati, T. Assessment and optimization of forest biomass supply chains from economic, social and environmental perspectives—A review of literature. *Renew. Sustain. Energy Rev.* **2014**, *36*, 62–73. [CrossRef]
3. Ferrarini, A.; Serra, P.; Almagro, M.; Trevisan, M.; Amaducci, S. Multiple ecosystem services provision and biomass logistics management in bioenergy buffers: A state-of-the-art review. *Renew. Sustain. Energy Rev.* **2017**, *73*, 277–290. [CrossRef]
4. Awudu, I.; Zhang, J. Uncertainties and sustainability concepts in biofuel supply chain management: A review. *Renew. Sustain. Energy Rev.* **2012**, *16*, 1359–1368. [CrossRef]
5. Wang, L.; Agyemang, S.A.; Amini, H.; Shahbazi, A. Mathematical modeling of production and biorefinery of energy crops. *Renew. Sustain. Energy Rev.* **2015**, *43*, 530–544. [CrossRef]

6. Van der Vorst, J.G.A.J.; Da Silva, C.; Trienekens, J.H. *Agro-Industrial Supply-Chain Management: Concepts and Applications*; FAO: Rome, Italy, 2007; Volume 17.
7. De Laporte, A.V.; Ripplinger, D.G. The effects of site selection, opportunity costs and transportation costs on bioethanol production. *Renew. Energy* **2019**, *131*, 73–82. [CrossRef]
8. Rodias, E.C.; Lampridi, M.; Sopegno, A.; Berruto, R.; Banias, G.; Bochtis, D.D.; Busato, P. Optimal energy performance on allocating energy crops. *Biosyst. Eng.* **2019**, *181*, 11–27. [CrossRef]
9. Kim, S.; Kim, S.; Kiniry, J.R. Two-phase simulation-based location-allocation optimization of biomass storage distribution. *Simul. Model. Pract. Theory* **2018**, *86*, 155–168. [CrossRef]
10. Perrin, A.; Wohlfahrt, J.; Morandi, F.; Østergård, H.; Flatberg, T.; De La Rua, C.; Bjørkvoll, T.; Gabrielle, B. Integrated design and sustainable assessment of innovative biomass supply chains: A case-study on miscanthus in France. *Appl. Energy* **2017**, *204*, 66–77. [CrossRef]
11. Sopegno, A.; Rodias, E.; Bochtis, D.; Busato, P.; Berruto, R.; Boero, V.; Sørensen, C.G. Model for Energy Analysis of Miscanthus Production and Transportation. *Energies* **2016**, *9*, 392. [CrossRef]
12. De Laporte, A.V.; Weersink, A.J.; McKenney, D.W. Effects of supply chain structure and biomass prices on bioenergy feedstock supply. *Appl. Energy* **2016**, *183*, 1053–1064. [CrossRef]
13. Strullu, L.; Ferchaud, F.; Yates, N.; Shield, I.; Beaudoin, N.; De Cortázar-Atauri, I.G.; Besnard, A.; Mary, B. Multisite Yield Gap Analysis of Miscanthus × giganteus Using the STICS Model. *BioEnergy Res.* **2015**, *8*, 1735–1749. [CrossRef]
14. Richter, G.M.; Agostini, F.; Barker, A.; Costomiris, D.; Qi, A. Assessing on-farm productivity of Miscanthus crops by combining soil mapping, yield modelling and remote sensing. *Biomass Bioenergy* **2016**, *85*, 252–261. [CrossRef]
15. Pari, L.; Curt, M.D.; Sánchez, J.; Santangelo, E. Economic and energy analysis of different systems for giant reed (*Arundo donax* L.) harvesting in Italy and Spain. *Ind. Crop. Prod.* **2016**, *84*, 176–188. [CrossRef]
16. Lozano-Moreno, J.A.; Maréchal, F. Biomass logistics and environmental impact modelling for sugar-ethanol production. *J. Clean. Prod.* **2019**, *210*, 317–324. [CrossRef]
17. Melendez, J.; Stuart, P.R. Systematic assessment of triticale-based biorefinery strategies: A biomass procurement strategy for economic success. *Biofuels Bioprod. Biorefining* **2018**, *12*, S21–S34. [CrossRef]
18. Scordia, D.; Testa, G.; Copani, V.; Patanè, C.; Cosentino, S.L. Lignocellulosic biomass production of Mediterranean wild accessions (*Oryzopsis miliacea*, *Cymbopogon hirtus*, *Sorghum halepense* and *Saccharum spontaneum*) in a semi-arid environment. *Field Crop. Res.* **2017**, *214*, 56–65. [CrossRef]
19. Busato, P.; Sopegno, A.; Berruto, R.; Bochtis, D.; Calvo, A. A Web-Based Tool for Energy Balance Estimation in Multiple-Crops Production Systems. *Sustainability* **2017**, *9*, 789. [CrossRef]
20. Rodias, E.; Berruto, R.; Bochtis, D.; Busato, P.; Sopegno, A. A Computational Tool for Comparative Energy Cost Analysis of Multiple-Crop Production Systems. *Energies* **2017**, *10*, 831. [CrossRef]
21. Wienforth, B.; Knieß, A.; Böttcher, U.; Herrmann, A.; Sieling, K.; Taube, F.; Kage, H. Evaluating Bioenergy Cropping Systems towards Productivity and Resource Use Efficiencies: An Analysis Based on Field Experiments and Simulation Modelling. *Agronomy* **2018**, *8*, 117. [CrossRef]
22. Bonner, I.; McNunn, G.; Muth, D.; Tyner, W.; Leirer, J.; Dakins, M. Development of integrated bioenergy production systems using precision conservation and multicriteria decision analysis techniques. *J. Soil Water Conserv.* **2016**, *71*, 182–193. [CrossRef]
23. Machado, K.S.; Seleme, R.; Maceno, M.M.; Zattar, I.C. Carbon footprint in the ethanol feedstocks cultivation—Agricultural CO_2 emission assessment. *Agric. Syst.* **2017**, *157*, 140–145. [CrossRef]
24. Fernando, A.L.; Costa, J.; Barbosa, B.; Monti, A.; Rettenmaier, N. Environmental impact assessment of perennial crops cultivation on marginal soils in the Mediterranean Region. *Biomass Bioenergy* **2018**, *111*, 174–186. [CrossRef]
25. Dragoni, F.; Di Nasso, N.N.O.; Tozzini, C.; Bonari, E.; Ragaglini, G. Aboveground Yield and Biomass Quality of Giant Reed (*Arundo donax* L.) as Affected by Harvest Time and Frequency. *BioEnergy Res.* **2015**, *8*, 1321–1331. [CrossRef]
26. Garofalo, P.; D'Andrea, L.; Vonella, A.V.; Rinaldi, M.; Palumbo, A.D. Energy performance and efficiency of two sugar crops for the biofuel supply chain. Perspectives for sustainable field management in southern Italy. *Energy* **2015**, *93*, 1548–1557. [CrossRef]

27. Schulz, V.; Munz, S.; Stolzenburg, K.; Hartung, J.; Weisenburger, S.; Mastel, K.; Möller, K.; Claupein, W.; Graeff-Hönninger, S. Biomass and Biogas Yield of Maize (*Zea mays* L.) Grown under Artificial Shading. *Agriculture* **2018**, *8*, 178. [CrossRef]
28. Rodias, E.; Berruto, R.; Busato, P.; Bochtis, D.; Sørensen, C.G.; Zhou, K. Energy Savings from Optimised In-Field Route Planning for Agricultural Machinery. *Sustainability* **2017**, *9*, 1956. [CrossRef]
29. Sahoo, K.; Mani, S.; Das, L.; Bettinger, P. GIS-based assessment of sustainable crop residues for optimal siting of biogas plants. *Biomass Bioenergy* **2018**, *110*, 63–74. [CrossRef]
30. Batidzirai, B.; Valk, M.; Wicke, B.; Junginger, M.; Daioglou, V.; Euler, W.; Faaij, A. Current and future technical, economic and environmental feasibility of maize and wheat residues supply for biomass energy application: Illustrated for South Africa. *Biomass Bioenergy* **2016**, *92*, 106–129. [CrossRef]
31. Mupondwa, E.; Li, X.; Tabil, L. Large-scale commercial production of cellulosic ethanol from agricultural residues: A case study of wheat straw in the Canadian Prairies. *Biofuels Bioprod. Biorefining* **2017**, *11*, 955–970. [CrossRef]
32. Manandhar, A.; Shah, A. Life cycle assessment of feedstock supply systems for cellulosic biorefineries using corn stover transported in conventional bale and densified pellet formats. *J. Clean. Prod.* **2017**, *166*, 601–614. [CrossRef]
33. Sahoo, K.; Hawkins, G.; Yao, X.; Samples, K.; Mani, S. GIS-based biomass assessment and supply logistics system for a sustainable biorefinery: A case study with cotton stalks in the Southeastern US. *Appl. Energy* **2016**, *182*, 260–273. [CrossRef]
34. Sahoo, K.; Mani, S. Engineering Economics of Cotton Stalk Supply Logistics Systems for Bioenergy Applications. *Trans. ASABE* **2016**, *59*, 737–747.
35. Santibañez-Aguilar, J.E.; Flores-Tlacuahuac, A.; Betancourt-Galvan, F.; Lozano-García, D.F.; Lozano, F.J. Facilities Location for Residual Biomass Production System Using Geographic Information System under Uncertainty. *ACS Sustain. Chem. Eng.* **2018**, *6*, 3331–3348. [CrossRef]
36. Rodríguez, R.; Gauthier-Maradei, P.; Escalante, H. Fuzzy spatial decision tool to rank suitable sites for allocation of bioenergy plants based on crop residue. *Biomass Bioenergy* **2017**, *100*, 17–30. [CrossRef]
37. Bao, X.; Yang, C.; Zhang, L. Location-route collaborative decision-making of a forestry and agricultural residue recycling network. *AGRO Food Ind. Hi-Tech.* **2016**, *27*, 125–130.
38. Gebresenbet, G.; Bosona, T.; Olsson, S.-O.; Garcia, D. Smart System for the Optimization of Logistics Performance of the Pruning Biomass Value Chain. *Appl. Sci.* **2018**, *8*, 1162. [CrossRef]
39. Estornell, J.; Ruiz, L.A.; Marti, B.V.; López-Cortés, I.; Salazar, D.; Fernández Sarría, A. Estimation of pruning biomass of olive trees using airborne discrete-return LiDAR data. *Biomass Bioenergy* **2015**, *81*, 315–321. [CrossRef]
40. Sánchez-García, S.; Athanassiadis, D.; Martínez-Alonso, C.; Tolosana, E.; Majada, J.; Canga, E. A GIS methodology for optimal location of a wood-fired power plant: Quantification of available woodfuel, supply chain costs and GHG emissions. *J. Clean. Prod.* **2017**, *157*, 201–212. [CrossRef]
41. Nilsson, D.; Larsolle, A.; Nordh, N.-E.; Hansson, P.-A. Dynamic modelling of cut-and-store systems for year-round deliveries of short rotation coppice willow. *Biosyst. Eng.* **2017**, *159*, 70–88. [CrossRef]
42. Schweier, J.; Becker, G.; Schnitzler, J.-P. Selected environmental impacts of the technical production of wood chips from poplar short rotation coppice on marginal land. *Biomass Bioenergy* **2016**, *85*, 235–242. [CrossRef]
43. Simões, D.; Dinardi, A.J.; Da Silva, M.R.; Silva, M. Investment Uncertainty Analysis in Eucalyptus Bole Biomass Production in Brazil. *Forests* **2018**, *9*, 384. [CrossRef]
44. Ebadian, M.; Shedden, M.E.; Webb, E.; Sokhansanj, S.; Eisenbies, M.; Volk, T.; Heavey, J.; Hallen, K. Impact of Parcel Size, Field Shape, Crop Yield, Storage Location, and Collection Equipment on the Performance of Single-Pass Cut-and-Chip Harvest System in Commercial Shrub Willow Fields. *BioEnergy Res.* **2018**, *11*, 364–381. [CrossRef]
45. Bisaglia, C.; Brambilla, M.; Cutini, M.; Bortolotti, A.; Rota, G.; Minuti, G.; Sargiani, R. Reusing Pruning Residues for Thermal Energy Production: A Mobile App to Match Biomass Availability with the Heating Energy Balance of Agro-Industrial Buildings. *Sustainability* **2018**, *10*, 4218. [CrossRef]
46. Krzyżaniak, M.; Stolarski, M.J.; Warmiński, K. Life cycle assessment of poplar production: Environmental impact of different soil enrichment methods. *J. Clean. Prod.* **2019**, *206*, 785–796. [CrossRef]

47. Miguel, G.S.; Corona, B.; Ruiz, D.; Landholm, D.; Laina, R.; Tolosana, E.; Sixto, H.; Cañellas, I. Environmental, energy and economic analysis of a biomass supply chain based on a poplar short rotation coppice in Spain. *J. Clean. Prod.* **2015**, *94*, 93–101. [CrossRef]
48. Manzone, M.; Calvo, A. Woodchip transportation: Climatic and congestion influence on productivity, energy and CO_2 emission of agricultural and industrial convoys. *Renew. Energy* **2017**, *108*, 250–259. [CrossRef]
49. Dufossé, K.; Drouet, J.-L.; Gabrielle, B. Agro-ecosystem modeling can aid in the optimization of biomass feedstock supply. *Environ. Model. Softw.* **2016**, *85*, 139–155. [CrossRef]
50. Nair, S.K.; Hartley, D.S.; Gardner, T.A.; McNunn, G.; Searcy, E.M. An Integrated Landscape Management Approach to Sustainable Bioenergy Production. *BioEnergy Res.* **2017**, *10*, 929–948. [CrossRef]
51. Busato, P.; Berruto, R. A web-based tool for biomass production systems. *Biosyst. Eng.* **2014**, *120*, 102–116. [CrossRef]
52. Lee, M.; Cho, S.; Kim, J. A comprehensive model for design and analysis of bioethanol production and supply strategies from lignocellulosic biomass. *Renew. Energy* **2017**, *112*, 247–259. [CrossRef]
53. Lin, T.; Wang, S.; Rodríguez, L.F.; Hu, H.; Liu, Y.Y. CyberGIS-enabled decision support platform for biomass supply chain optimization. *Environ. Model. Softw.* **2015**, *70*, 138–148. [CrossRef]
54. Kaliyan, N.; Morey, R.V.; Tiffany, D.G. Economic and Environmental Analysis for Corn Stover and Switchgrass Supply Logistics. *BioEnergy Res.* **2015**, *8*, 1433–1448. [CrossRef]
55. Pavlou, D.; Orfanou, A.; Busato, P.; Berruto, R.; Sørensen, C.; Bochtis, D. Functional modeling for green biomass supply chains. *Comput. Electron. Agric.* **2016**, *122*, 29–40. [CrossRef]
56. Subhashree, S.N.; Igathinathane, C.; Bora, G.C.; Ripplinger, D.; Backer, L. Optimized location of biomass bales stack for efficient logistics. *Biomass Bioenergy* **2017**, *96*, 130–141. [CrossRef]
57. Igathinathane, C.; Tumuluru, J.; Keshwani, D.; Schmer, M.; Archer, D.; Liebig, M.; Halvorson, J.; Hendrickson, J.; Kronberg, S. Biomass bale stack and field outlet locations assessment for efficient infield logistics. *Biomass Bioenergy* **2016**, *91*, 217–226. [CrossRef]
58. Edwards, G.; Sørensen, C.G.; Bochtis, D.D.; Munkholm, L.J. Optimised schedules for sequential agricultural operations using a Tabu Search method. *Comput. Electron. Agric.* **2015**, *117*, 102–113. [CrossRef]
59. Liakos, K.G.; Busato, P.; Moshou, D.; Pearson, S.; Bochtis, D. Machine Learning in Agriculture: A Review. *Sensors* **2018**, *18*, 2674. [CrossRef]

energies

MDPI

Article

Energy Footprint of Mechanized Agricultural Operations

Maria Lampridi [1,2,*], Dimitrios Kateris [1], Claus Grøn Sørensen [3] and Dionysis Bochtis [1]

1 Center for Research and Technology Hellas (CERTH), Institute for Bio-Economy and Agri-technology (IBO), 6th km Charilaou-Thermi Rd., 57001 Thessaloniki, Greece; d.kateris@certh.gr (D.K.); d.bochtis@certh.gr (D.B.)

2 School of Agriculture, Faculty of Agriculture, Forestry and Natural Environment, Aristotle University of Thessaloniki, 54124 Thessaloniki, Greece

3 Department of Engineering, Aarhus University, Blichers Allé 20, P.O. Box 50, 8830 Tjele, Denmark; claus.soerensen@eng.au.dk

* Correspondence: m.lampridi@certh.gr; Tel.: +30-2311-257-650

Received: 19 December 2019; Accepted: 7 February 2020; Published: 10 February 2020

Abstract: The calculation of the energy cost of a cultivation is a determining factor in the overall assessment of agricultural sustainability. Most studies mainly examine the entire life cycle of the operation, considering reference values and reference databases for the determination of the machinery contribution to the overall energy balance. This study presents a modelling methodology for the precise calculation of the energy cost of performing an agricultural operation. The model incorporates operational management into the calculation, while simultaneously considering the commercially available machinery (implements and tractors). As a case study, the operation of tillage was used considering both primary and secondary tillage (moldboard plow and field cultivator, respectively). The results show the importance of including specific operation parameters and the available machinery as part of determining the accurate total energy consumption, even though the field size and available time do not have a significant effect.

Keywords: agricultural operations; energy use; assessment tool; workability; machinery

1. Introduction

The concept of agricultural sustainability has been a highly debated subject in recent years, not only because of its importance with respect to evaluating the agri-food system, but mostly because of its multivariate nature [1]. In fact, there is a large number of parameters that are considered with respect to the primary production of agricultural products [2]. These parameters include not only the relevant material flows (e.g., water, fuel, fertilizers, seeds, and crops), but also the means of production, which include labor and the use of agricultural machinery along with the inclusion of operation management [3]. The integration of the individual evaluation of each of those parameters with respect to their contribution to sustainability concerns more and more researchers in recent years. However, agricultural sustainability is approached in a variety of ways in the literature, examining, individually or in combination, economic, environmental, and social indicators [4]. These indicators attempt to assess the impacts of agriculture in a quantitative or a qualitative manner. Several indicators as well as indicator indexes and frameworks have been proposed in the literature with respect to each dimension of sustainability [5]. The most frequently used economic indicators include profitability, income, efficiency, productivity and yield, while the societal indicators include the education, health risk, employment, operational difficulties and access to health [6–9]. With respect to the environmental indicators, the most popular are energy use, water footprint, soil quality, biodiversity, and greenhouse gas (GHG) emissions [10–13].

In particular, energy cost (or energy consumption or energy input) is one of the most examined agricultural sustainability indicators of field operations [14]. Energy consumption is an important agricultural environmental indicator and serves as a basis for the calculation of several composite indicators related to energy such as energy efficiency and energy balance [14,15]. The research focus during recent years was on the assessment and calculation of the energy footprint for the entire crop production, providing approaches for improvement in terms of environmental impact measures, through the minimization of agricultural inputs such as water use, fertilizers, and plant protection products [14–17]. In all of the cases, the goal is to assess or minimize the environmental burden of agricultural operations, while maintaining the quality and quantity of production [13]. For these assessments, the calculation of the energy consumption was mostly based on reference values and databases, while the models used do not always consider the commercially available equipment and mostly focus on material consumption [18]. Additionally, the effect of machinery selection or machinery dimensioning (considering, for example, the time availability, that is, the workability of the system at hand) is not considered during the assessment, while there are only a few studies examining the embodied energy of agricultural machinery [19–21] and the energy cost of machinery when performing a specific task [22]. It is also worth noting that, with respect to the selection of the required machinery, mostly financial methods are employed that consider the economic cost of operations [23,24]. However, with the advancement of technology it is important to examine the energy contribution of the machinery, especially when the introduction of new equipment is assessed [25–27].

Considering the above and attempting to fill the identified research gap in the field of agricultural machinery energy assessment, the present study attempts to incorporate "machinery dimensioning notion" in the calculation of the energy costs. The ultimate goal is to assess the influence of the workability that determines the available time window for performing a specific field operation on the resulting energy consumption for this operation. Furthermore, the machinery used is assessed, examining the contribution of the standardized-commercial machinery in the overall energy requirements. For that reason, a computational model was designed and implemented in order to examine the machinery energy cost of performing an agricultural operation. The model selects the required machinery to perform the operation under given workability constraints [23], choosing from a collection of commercially available implements and tractors, and considering a series of operational parameters including the soil type, the operation depth, and the field area to be processed.

In the following sections, the methods and the detailed design of the model are presented followed by results from two case studies. The examined operations concern the soil preparation and, more specifically, light soil cultivation with a field cultivator and primary tillage with a moldboard plow.

2. Materials and Methods

2.1. Energy Cost of Agricultural Operations

The execution of an agricultural operation requires both direct and indirect inputs. To that end, the employed approach in this study borrows from the well-established methodology of financial cost calculation [28], and allocates the total energy required to perform an operation to direct and indirect energy (Figure 1). The indirect energy concerns the energy assessment of inputs that are not related to the operation performed but concern the acquisition and the ownership of the equipment. The direct energy involves the energy assessment of the direct inputs related to agricultural operations.

On the basis of the above, the indirect energy is differentiated between the machinery and repair and maintenance embodied energy ($MJ \cdot kg^{-1}$), as well as the housing embodied energy ($MJ \cdot m^{-2} \cdot y^{-1}$). The machinery embodied energy refers to the energy required to manufacture the agricultural equipment [20,29]. The embodied energy of repair and maintenance can be either expressed as a percentage of the manufacturing energy requirements [30,31] or can be estimated in absolute numbers, as attempted by Mantoam et al. [32]. The housing embodied energy represents the energy contribution of all the building and infrastructure necessary for the storage of the equipment [30].

It should be noted that these denoted parameters are usually calculated for the entire life cycle of the machinery or building [33]. In this study, the energy contribution from manufacturing, repair and maintenance, and housing is reduced to the duration of the operation performed considering as related to the total life of the equipment.

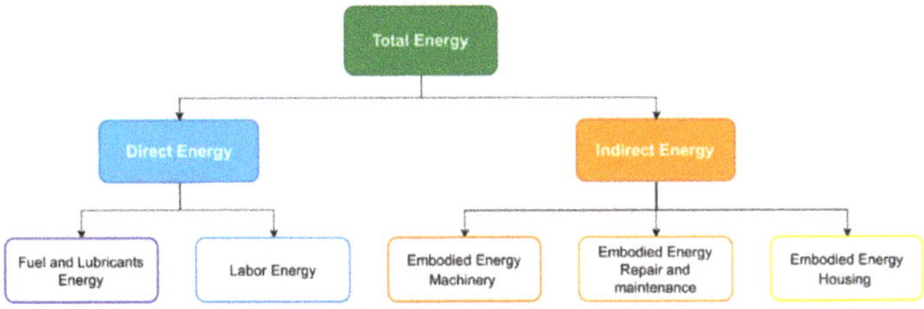

Figure 1. Energy costs of agricultural operations.

The operation energy is divided into the fuel and lubricant energy ($MJ \cdot l^{-1}$), and the embodied energy of labor ($MJ \cdot h^{-1}$). The energy of fuels and lubricants is related to the fuel and lubricant consumption as calculated based on the ASABE (American Society of Agricultural and Biological Engineer) standards [34]. The embodied energy of labor expresses the energy consumed by the workers performing the operation [35]. Several approaches have been proposed in the literature regarding the estimation of the embodied energy of labor differing in the system boundaries and operations considered [16,36–39].

2.2. Energy Estimation Model

In order to calculate the energy cost of performing an agricultural operation, the model presented in Figure 2 was used. The calculation consists of two distinct stages, namely the machinery selection and the calculation of the energy costs. For the machinery selection stage, in order to estimate the effect of standardized machinery on the overall energy consumption, two models were considered, namely the continuous and the discrete model, while the energy cost calculation is the same for both models. The continuous model calculates the energy consumption of a theoretical machine system without correction to standard commercial machinery, while the discrete model considers the commercially available implements and tractors for the determination of the energy costs. More specifically, the necessary machine system (single or multiple if it is required) to perform the operation is automatically calculated by the model, considering both operational parameters and the machinery available. The calculation process of the discrete model is presented in detail below, however, the basic equations are standard for both models. Nevertheless, it should be highlighted that, in the continuous model, only a single machine system is considered.

During the machinery selection stage, first the available time t_{av} (h) to perform an agricultural operation is calculated (Equation (1)):

$$t_{av} = t_{wp} \cdot t_{wt} \cdot W_{coef} \tag{1}$$

where t_{wp} (d) is the working period in days, t_{wt} $\left(h \cdot d^{-1}\right)$ is the working time, and W_{coef} $(-)$ is the workability coefficient [40]. Given a specific available time, the theoretical area capacity C_{th} $\left(m^2 \cdot h^{-1}\right)$ is calculated (Equation (2)):

$$C_{th} = \frac{A}{t_{av}} \tag{2}$$

where A $\left(m^2\right)$ is the size of the field under examination. The theoretical capacity determines, as a sequence, the theoretical minimum width w_{th} (m) (Equation (3)) that must be applied in order to complete the operation on time.

$$w_{th} = \frac{C_{th}}{E_f \cdot s} \tag{3}$$

where E_f (%) stands for the efficiency of the operation performed, while s $\left(km \cdot h^{-1}\right)$ is the operating speed.

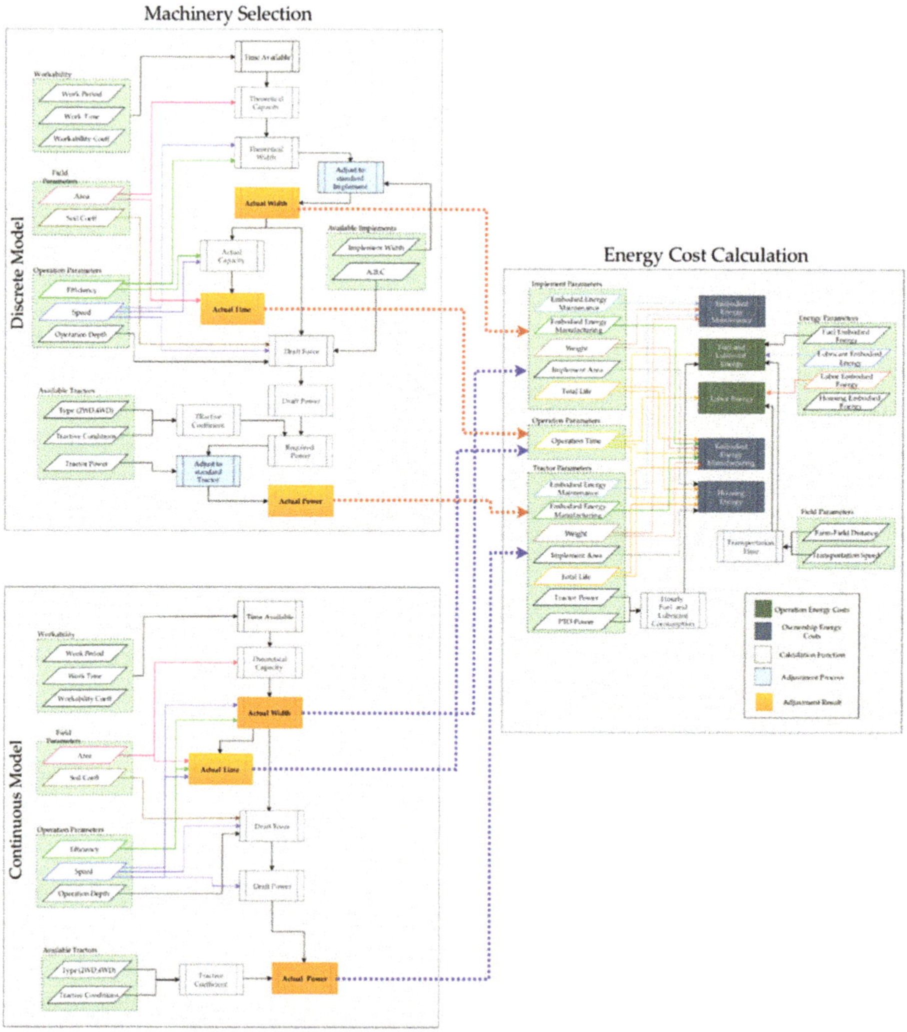

Figure 2. Calculation model.

Then, according to the discrete model, the theoretical width is adjusted based on the commercially available implements. When the theoretical width is smaller than the maximum width available, then the width selected is the one that is immediately bigger than the theoretical width. When the theoretical width is larger than the maximum width available, it means that more than one implements are required to perform the operation on time. In that case, the maximum width is selected (as many times as required), while the remaining width is adjusted to the next bigger available width. In should be noted that, in this step, the actual number of machines required to perform the operation is determined. Considering the above, the actual width w_{act} (m) of the operation performed is formulated and, as a consequence, the actual area capacity is calculated by Equation (4) as follows:

$$C_{act} = w_{act} \cdot s \cdot E_f \tag{4}$$

Finally, the actual duration of the operation is calculated (Equation (5)):

$$t_{act} = \frac{A}{C_{act}} \tag{5}$$

In the case of the continuous model where a correction to standard machines is not performed, the actual width and actual capacity are the same as the theoretical width and capacity and actual time coincides with the time available to perform the operation. Once the required implements have been determined, the calculation of the required tractors follows. The implement draft D_{draft} (N) is calculated according to the ASABE standards [41] by Equation (6):

$$D_{draft} = F_i \cdot \left[A + B \cdot s + C \cdot s^2\right] \cdot w_{act} \cdot O_{depth} \tag{6}$$

where F_i (−) is the soil texture adjustment parameter; A, B, and C are machine specific parameters; and O_{depth} (cm) is the operation depth. Considering the above, the draft power (kW) is calculated as follows [23]:

$$P_{draft} = \frac{D_{draft} \cdot s}{1000} \tag{7}$$

Finally, the total power required $P_{required}$ (kW) is calculated from Equation (8) as follows:

$$P_{required} = \frac{P_{draft}}{Tr_{coef} \cdot 0.83} \tag{8}$$

where Tr_{coef} (−) stands for the tractive coefficient [42], while the coefficient 0.83 is used for the Power Take Off (PTO) to gross-flywheel conversion [34]. The calculated $P_{required}$ (kW) corresponds to the minimum power required to perform the operation considering the necessary implements as they were defined in the previous steps. However, commercial tractors are available with a standard power capacity for each model, as defined by the manufacturer. To that end, the required power is adjusted to the next bigger available. In the case of the continuous model, the required power coincides with the actual power because there is not a correction to standard tractors. The machinery selection stage provides the list of tractors-implements that are necessary for the completion of the procedure in the given time available for the discrete model and the single tractor and implement system for the continuous model. For each tractor-implement set, the energy costs of performing the operation are calculated.

Regarding the operation parameters calculations, the hourly fuel consumption F_{con} $\left(l \cdot h^{-1}\right)$ is calculated from Equation (9) (for diesel) as follows [34]:

$$F_{con} = \left(2.64 \cdot P_{ratio} + 3.91 - 0.203 \cdot \sqrt{738 \cdot P_{ratio} + 173)}\right) \cdot P_{required} \tag{9}$$

where P_{ratio} expresses the ratio of equivalent PTO power necessary for an operation to that maximum available from the PTO [34]. The relevant lubricant consumption $L_{con}\left(l \cdot h^{-1}\right)$ is calculated in Equation (10) (for diesel) [34].

$$L_{con} = 0.00059 \cdot P_{required} + 0.02169 \tag{10}$$

Lastly, it is important to note that, for the calculation of the operation energy costs, the transportation duration is also considered along with the operation duration. The resulting per hectare total energy En_{Tot} (MJ) is calculated as the sum of the direct En_{Direct} and indirect energy $En_{Indirect}$, as presented in Equation (11):

$$En_{Tot} = En_{Direct} + En_{Indirect} \tag{11}$$

The direct energy En_{Direct} (MJ) is calculated by Equation (12) as follows:

$$En_{Direct} = En_{Fuel} + En_{Lubr} + En_{Labor} \tag{12}$$

where En_{Fuel} (MJ) is the total fuel energy; En_{Lubr} is the total lubricant energy and En_{Labor} is the total labor energy expressed in Equations (13)–(15):

$$En_{Fuel} = F_{con} \cdot t_{act} \cdot Emb_{Fuel} \tag{13}$$

$$En_{Lubr} = L_{con} \cdot t_{act} \cdot Emb_{Lubr} \tag{14}$$

$$En_{Labor} = t_{act} \cdot Emb_{Labor} \tag{15}$$

where $Emb_{Fuel}\left(MJ \cdot l^{-1}\right)$ and $Emb_{Lubr}\left(MJ \cdot l^{-1}\right)$ are the embodied energy in the fuels and lubricants and Emb_{Labor} $(MJ \cdot h^{-1})$ is the per hour labor energy. With respect to the indirect energy consumption, Equation (16) applies as follows:

$$En_{Indirect} = En_{Manufacturing} + En_{RM} + En_{Housing} \tag{16}$$

The embodied energy for the manufacturing $En_{Manufacturing}$ (MJ) of the equipment (tractor and implement) is expressed in Equation (17), reduced to total hours of the operation:

$$En_{Manufacturing} = \frac{Emb_{Manufturing} \cdot M}{h_{life}} \cdot t_{act} \tag{17}$$

where $Emb_{Manufacturing}\left(MJ \cdot kg^{-1}\right)$ is the embodied energy; M (kg) is the mass of the tractor and the implement, respectively and h_{life} (h) is the total life hours of the machinery. The repair and maintenance embodied energy En_{RM} (MJ), as already mentioned, is expressed as a percentage of the manufacturing energy requirements RM_{Rate} (%) (Equation (18)):

$$En_{RM} = En_{Manufacturing} \cdot RM_{Rate} \tag{18}$$

Lastly, the housing energy $En_{Housing}$ (MJ) is calculated by Equation (19) reduced to total hours of the operation as follows:

$$En_{Housing} = \frac{Emb_{Housing} \cdot A_{cover}}{h_{life}} \cdot t_{act} \tag{19}$$

In that case, $Emb_{Housing}\left(MJ \cdot m^{-2} \cdot y^{-1}\right)$ is the embodied energy of the facility required to house the equipment and $A_{cover}\left(m^2\right)$ is the area cover of each the machines (tractor or implement).

3. Case Study Description

The methodology described in the previous section was used to investigate the energy cost of performing two different tillage operations, considering different field areas and the available time window. The examined operations are secondary tillage performed by a field cultivator and deep tillage performed by a moldboard plow. Table 1 presents the operation parameters considered according to the ASABE standards [34]. The economic and total life is assumed to be identical for both implements. However, the operating speed differs with respect to the different operations because it depends on the operation depth, which is determined based on the general cultivation practices [43].

Table 1. Operation input parameters.

	Units	Moldboard Plow	Field Cultivator
Implement Parameters A, B, and C [1]	-	652-0-5.1	32-1.9-0
Economic Life [1]	y	15	15
Total Life [1]	h	2000	2000
Efficiency [1]	%	80	85
Operating Speed [1]	$km \cdot h^{-1}$	7	8
Operating Depth [2]	cm	30	15
Soil Coefficient (Medium Soil) [1]	-	0.7	0.85
Implement Width Range *	m	1.1–4.4	4–6

[1] [34], [2] [43] * Commercial values.

As mentioned in the methodology, the estimated required implement width and the tractor power are adjusted to fit with the next bigger commercially available one. To that effect, feature values of commercially available machinery were collected for the calculations as they are provided by the manufactures. These specifications include the width, weight and dimensions of the implements and the power, weight, and dimensions of the tractors. All the tractors (with power ranging from 37.4 kW to 456 kW) that were inserted in the model are 4WD, while the fuel type is diesel in all the cases examined. Regarding the calculation of the transportation energy cost, it is assumed that each field is located at a distance of 1 km from the farm. The tractor travels at 40 $km \cdot h^{-1}$ and performs each trip once in order to complete the operation and return to the farm.

The embodied energy parameters considered for the model calculation are presented in Table 2. Various methodologies have been proposed for the calculation of the manufactured embodied energy of farm machinery; however, for the purposes of this study, the approach of Kitani et al. was selected [35]. With respect to the embodied energy of the maintenance of the machinery, it is calculated as a percentage of the respective manufacturing energy according to the approach of Aguilera et al. [30]. The approach of Aguilera et al. was also used for the estimation of the housing embodied energy, considering the average value per year and per spatial coverage of the service building. All the parameters presented in this section represent the input parameters of the calculation process of Figure 2, which are highlighted in the green boxes.

Table 2. Energy parameters.

Energy Parameter	Units	Value
Tractor Embodied Energy (Manufacturing) [1]	$MJ \cdot kg^{-1}$	138
Implement Embodied Energy (Manufacturing) [1]	$MJ \cdot kg^{-1}$	180
Tractor Embodied Energy (Maintenance) [2]	%	45
Implement Embodied Energy (Maintenance) [2]	%	30
Labor Embodied Energy [3]	$MJ \cdot h^{-1}$	2.2
Fuel Embodied Energy (Diesel) [1]	$MJ \cdot l^{-1}$	47.8
Lubricant Embodied Energy [4]	$MJ \cdot l^{-1}$	46
Housing Embodied Energy [2]	$MJ \cdot m^{-2} \cdot y^{-1}$	21

[1] [33], [2] [30], [3] [16], [4] [44].

4. Results

This section presents the results of the implementation of the energy calculation tool with respect to the estimation of the machinery energy cost of performing three different tillage operations. Two different assessment models were examined, namely the discrete and the continuous model. The discrete model refers to the calculation of the required energy using the adjustment to standard commercial implements and tractors according to the process presented in the methodology. The continuous model refers to the theoretical energy consumption as it would be estimated without the adjustment to standard implement and standard values. The machinery selected in the latter case would have the exact performance characteristics with the estimated requirements to perform the operation. Furthermore, in that case, there are no maximum available values, indicating that the operation is performed with one set of tractor–implement, irrespective of the width and power necessary to operate within the time window available, which coincides with the actual operation time. The figures below present the total energy consumption, the total machine power, the number of machines, as well as the total operating width with reference to the time available and the cultivation area.

4.1. Field Cultivator

With respect to the secondary tillage with a field cultivator, Figure 3 presents the total energy cost, the total machine power, the number of machines, and the total operating width for cultivated areas of 100 ha in relation to the available time to perform the operation. The energy consumption for the specific area varies around the average value of 234.99 MJ $\cdot$ ha^{-1}. The tool selects the required machinery system to perform the operation within the available time window. As expected, the number of machines required to complete the operation decreases as the available time increases. For example, three machines are required to complete the operation in 10 h and two machines when the available time exceeds 13 h. After 25 h, only one machine is required to complete the operation and, as a result, only the required width decreases. The total energy consumed shows an upward trend owing to increased total operation time until the 37 h available time. At this point, only the smallest implement is used to complete the operation, and the time of operation does not change further with the energy consumed, being the same irrespective of the time available.

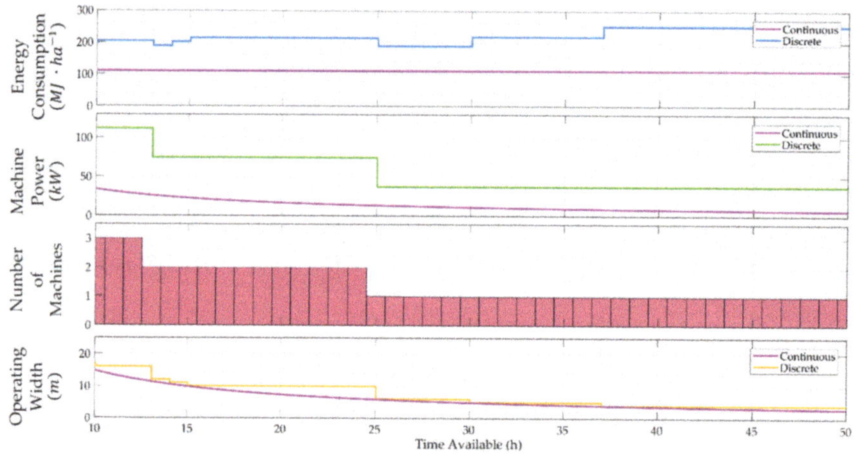

Figure 3. Energy consumption, machine power, number of machines, and operating width for a cultivated area of 100 ha (field cultivator).

The discrete model represents the realistic approach for the calculation of the energy required to perform an operation. However, as presented in Figure 3, it is worth comparing it to the continuous model. In the case of field cultivator, the continuous model predicts an average energy cost of 111.83 MJ $\cdot$ ha^{-1} for the area of 100 ha, which is 2.1 times smaller than the respective discrete. The tractor power and the implement width that are actually used are significantly larger than those that are ideally required, resulting in increased energy consumption.

Figures 4 and 5 present the energy cost and machinery requirements for completing the operation in 30 h and 50 h, respectively, in relation to the cultivated area. Little available time leads to a rapid increase of machinery needed, while in the case of more time availability, a second machine is required for fields larger than 200 ha. Considering that, for the field cultivator three, different implement options were available (4 m, 5 m, and 6 m), the operating width increases after an area of 135 ha for an available time of 50 h. The energy consumed shows a downwards trend when increasing the cultivated area, with average values of 218.94 MJ $\cdot$ ha^{-1} for 30 h and 228.98 MJ $\cdot$ ha^{-1} for 50 h available time. The average energy consumed increases when increasing the available time because the operating time increases as well. In the case of 50 h available, the energy consumed per hectare remains constant for fields between 2 and 135 ha with the given implements, as the smallest implement can cover this area without requiring a second one to perform the operation timely.

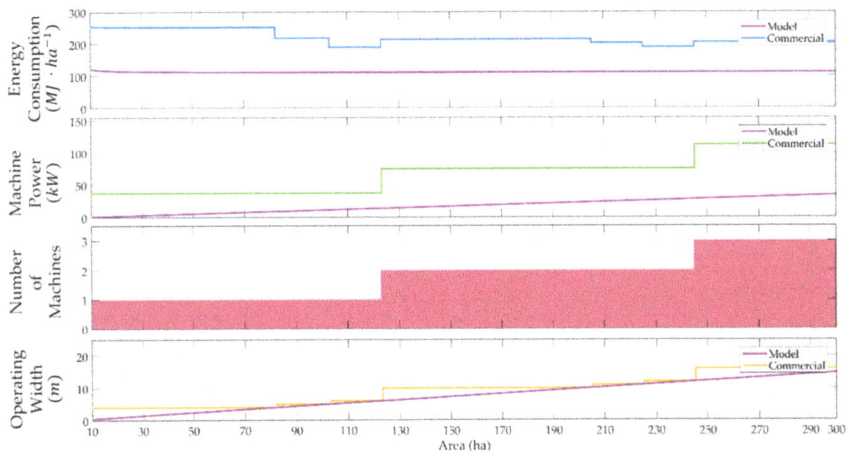

Figure 4. Energy consumption, machine power, number of machines, and operating width for 30 h time available (field cultivator).

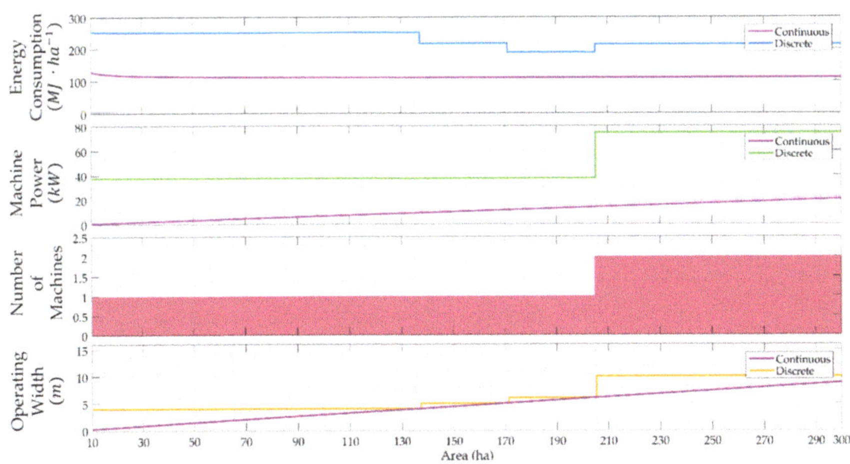

Figure 5. Energy consumption, machine power, number of machines, and operating width for 50 h time available (field cultivator).

In order to gain full insight regarding the formulation of the model, Figure 6 presents the results for the time available and the area, ranging from 10 h to 100 h and from 10 ha to 300 ha, respectively. The discrete model estimates a higher energy consumption per ha than the continuous model. The only exception is the area of 1 ha, where the continuous model is more energy consuming than the discrete model after approximately 40 h of time available owing to the very small stipulated operating width. However, as such a small operating width is not realistic (approximately 0.015 m for 1 ha and 100 h available time), energy values corresponding to areas below 10 ha were discarded from Figure 6. The total average energy cost is 226.51 $MJ \cdot ha^{-1}$ for the discrete model, which is 1.98 times higher than the continuous model. For the field cultivator, 69.4% of energy consumption derives from the operation of the machinery, while 30.6% is attributed to its ownership.

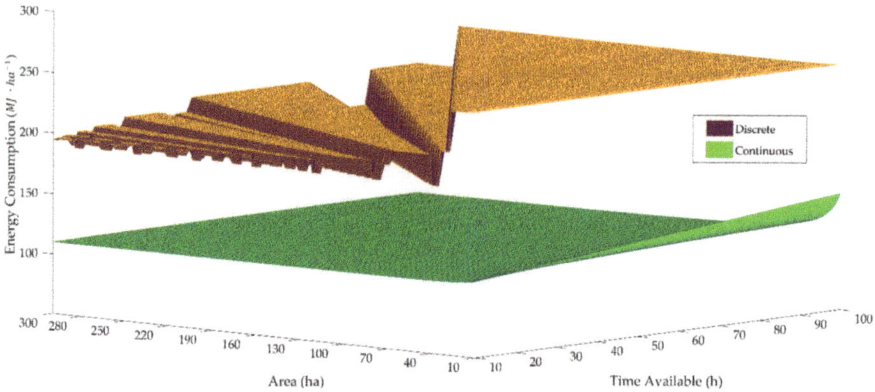

Figure 6. Energy consumption model for the field cultivator.

4.2. Moldboard Plow

Figure 7 presents the total energy consumption of performing primary tillage with a moldboard plow in a field of 100ha in relation to the time available to perform the operation. The average energy consumption for the 100 ha is 1505.82 MJ $\cdot$ ha^{-1} for the discrete model. It is observed that the continuous model shows higher energy consumption, approximately 1.06 times, than the discrete model demonstrating the effect of the operation depth and type in the overall energy costs.

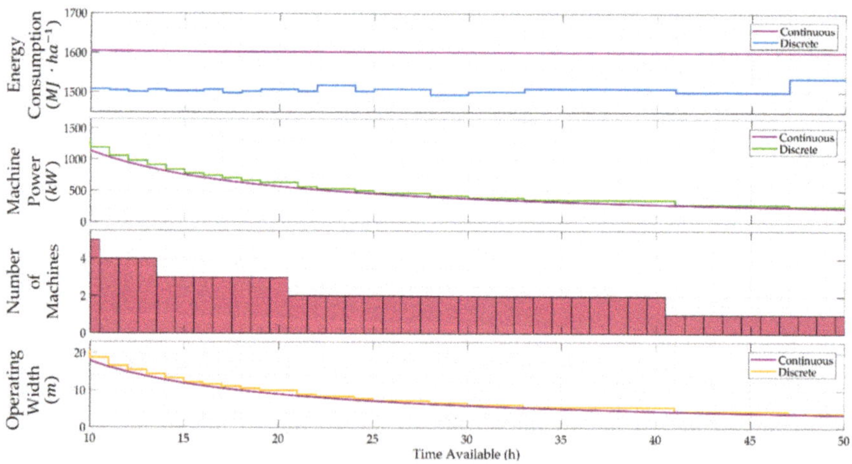

Figure 7. Energy consumption, machine power, number of machines, and operating width for a cultivated area of 100 ha (moldboard plow).

The required width and power to complete the operation on time do not differ significantly between the discrete and the continuous model. This can be attributed to the large energy requirements of the operation, as well as to the availability of implements. In the case of the moldboard plow, the commercially available width of the implements presented more subdivisions in the range of 1.1–4.4 m, providing the tool with more options during the machinery selection stage.

Figure 8 presents the evolution of the model with respect to the field area, considering time available of 50 h. For field areas up to 30 ha, the operation can be performed with one tractor carrying the minimum available implement (1.1 m). From this point, the implement width is further increased until it reaches the maximum available of 4.4 m. Then, with a further increase in the area, a second machine is needed to perform the operation timely. The width and power increasing pattern is the same for the second machinery as previously described until the point when a third machine is required. The average energy cost, in the case of 50 h of available time, is 1508.61 MJ · ha^{-1}.

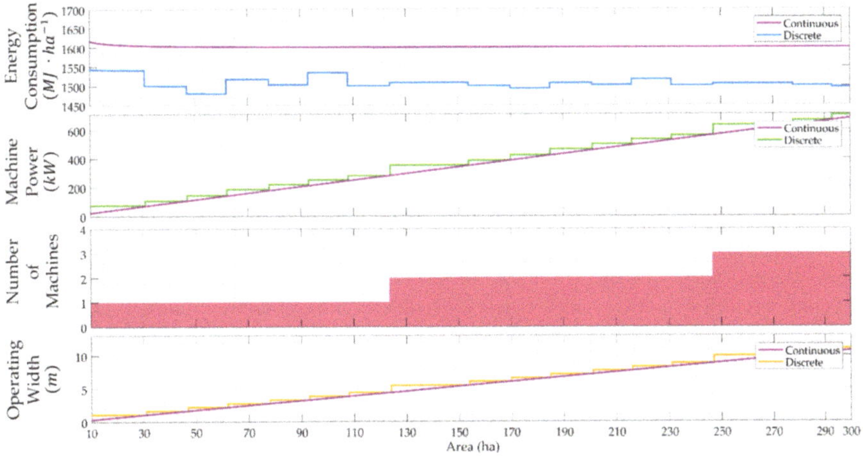

Figure 8. Energy consumption, machine power, number of machines, and operating width for 50 h time available (field cultivator).

The total energy consumption of the discrete and continuous model for moldboard plowing in a wide range of area and available time is presented in Figure 9. The continuous model has an average energy cost of 1604.95 MJ · ha^{-1}, which is 1.06 times higher than the discrete model, which presents an average energy cost per hectare of 1510.37 MJ · ha^{-1}. This finding highlights the importance of time when performing an agricultural operation of high operation energy requirements. For moldboard plowing, 90.4% of the energy consumed is attributed to operation, while only 9.6% of the energy consumed refers to the energy ownership costs. As a consequence, the adjustment to the closest bigger available implement resulted in a reduction of the total energy, as the actual time required to perform the operation decreased with the increased implements' width.

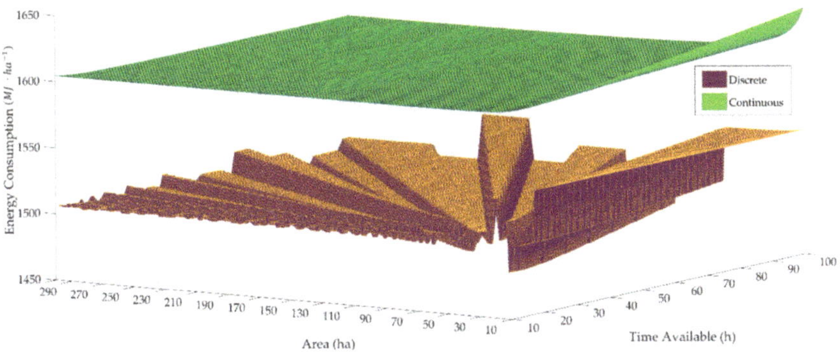

Figure 9. Energy consumption model for moldboard plow.

5. Discussion

In the previous section, the energy cost of performing secondary tillage with a field cultivator and primary tillage with a moldboard plow was presented. The results concerned tillage on medium soil with an operation depth of 15 cm and 30 cm, respectively. Tillage with a moldboard plow was calculated to be 6.67 times more energy consuming than field cultivation, which is an outcome that is expected considering the depth and the nature of the operation. For the field cultivator, the indirect energy costs account for almost 1/3 of the total energy consumed, while for the moldboard plow, only for 1/10. The direct energy cost is the determining factor of the total energy consumption. However, it is worth noting that its contribution to the total energy consumed increases as the energy intensity of the operation increases. Nevertheless, it should be stated that the available time to perform an operation and the different field areas do not seem to strongly affect the total energy per hectare.

Regarding machinery selection, the method employed focuses on the available time for field operations rather than the assessment of the financial cost of the equipment. In that light, the dimensioning of the implement requires sufficient power in the tractor as a prerequisite for the execution of the operation on time. The main disadvantage of the method is the fact that the operating economic costs are not considered during the selection stage. Additionally, the tool is designed in order to facilitate machinery selection, taking into account technical conditions rather than economic ones. As a consequence, a selection of the machinery system by applying this method can lead to non-optimum solutions in terms of economic criteria such as minimum operating or ownership cost. Furthermore, it is of limited use in cases where the equipment has already been purchased.

The results indicate that the proper allocation of tools and tractors and the use of the appropriate combination of machinery in each operation can affect the total energy consumed. In the case of light soil cultivation (field cultivator), the total energy consumed is higher than that expected from model calculations owing to the use of larger implements than those required to perform the operation on time. On the other hand, in the case of the moldboard plow, the use of a larger implement than the one required to perform the operation on time leads to a decreased energy consumption, which is justified when considering the contribution of the operation cost to the total energy cost of this operation. Nevertheless, the availability and the subdivision of the implement width strongly affect the diversion from the model consumption calculated for the examined cases.

Another factor that was examined was the sensitivity of the energy consumed with respect to the type of soil and the operation depth. In that end, the calculations were performed for different depths (plus 5 cm and minus 5 cm from the original example), as presented in Figure 10, and soil types (fine and coarse in addition to medium soil), as presented in Figure 11. For the field cultivator, it was found that the operation depth and the type of soil do not affect the total energy consumed in the case of the discrete model. This is attributed to the fact that the minimum available commercial machinery covers the requirements of the operation, irrespective of the type of soil or the depth.

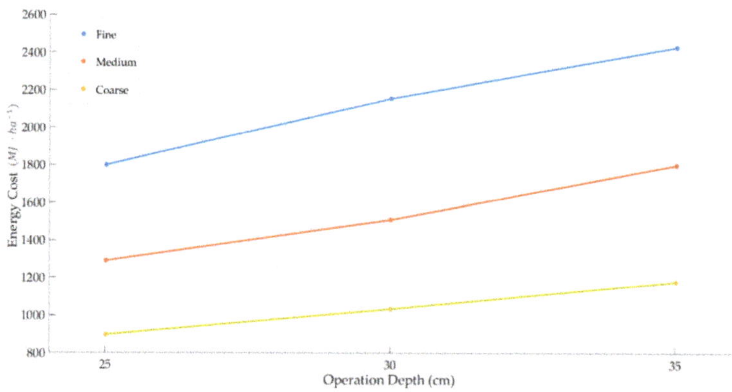

Figure 10. Moldboard plow energy consumption for various soil types and operation depths.

However, in the case of moldboard plowing, both the operation depth and the type of soil strongly influence the total energy consumed. As presented in Figure 10, the energy consumed for deep tillage in fine soil is almost two times higher than the that for coarse soil in the case of an operation depth of 25 cm. Additionally, in the case of fine soil, the energy for performing the operation at an operation depth of 35 cm is 2429.54 MJ $\cdot$ ha^{-1}, 1.35 times higher than when performing it at a depth of 25 cm (1800.88 MJ $\cdot$ ha^{-1}).

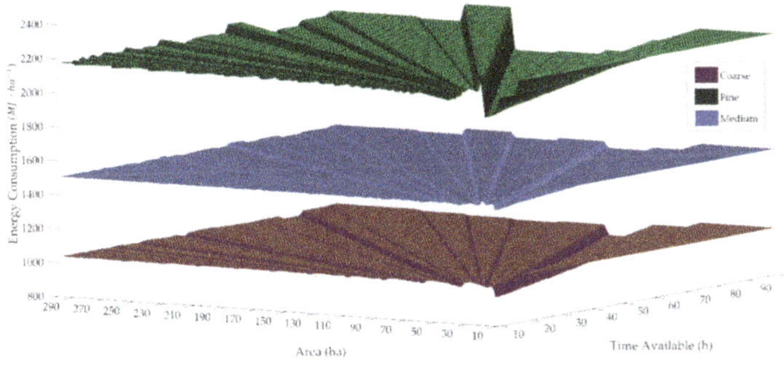

Figure 11. Moldboard plow energy consumption discrete model for various soil types.

6. Conclusions

The assessment and management of agricultural production is a multivariate problem. Thereby, the introduction and use of operation management tools can contribute towards the effective management of agricultural production and the quantification of its environmental impact. Existing tools and methodologies in most cases examine the entire life cycle of the cultivation or focus on the use of plant protection products and fertilizers owing to their proven severe adverse impact compared with the other agricultural inputs. Nevertheless, the quantification of the contribution of machinery to the total energy consumed is important and constitutes an essential first step for the evaluation of alternative technologies and tools. In fact, the results highlight the benefits of optimal allocation of tools, especially in the task of light soil cultivation, where the energy consumed using the commercially available machinery (226.51 MJ $\cdot$ ha^{-1}) is almost double the optimal energy expressed by the continuous model (113.88 MJ $\cdot$ ha^{-1}).

This paper attempted to examine and set the base of evaluating the energy consumption considering various operational parameters coupled with the commercial availability of implements and tractors. The model selects (from a collection of available implements and tractors) the required machinery to perform the operation in a given time, while evaluating the resulting energy cost. Even though the available time and the cultivated area do not seem to strongly affect the energy consumption per hectare, the commercially available implements highly affect the total energy cost. From the examined case studies, the effect of using unsuitable equipment was highlighted, indicating a promising potential for further reduction in the energy consumed from the machinery. Towards this direction, future work incorporating the consideration of financial parameters during the machinery selection is required in order to assess the economic and environmental performance of the machinery in an integrated manner.

Author Contributions: Conceptualization, M.L., D.K., C.G.S., and D.B.; data curation, D.K. and D.B; formal analysis, D.K.; investigation, M.P. and D.K.; methodology, M.L. and D.K.; software, M.L.; supervision, D.B.; validation, D.K., C.G.S., and D.B.; writing—original draft, M.L.; writing—review and editing, D.K., C.G.S., and D.B. All authors have read and agreed to the published version of the manuscript.

Funding: The work was supported by the project "Research Synergy to address major challenges in the nexus: energy-environment-agricultural production (Food, Water, Materials)" NEXUS, funded by the Greek Secretariat for Research and Technology (GSRT), Pr. No. MIS 5002496.

Conflicts of Interest: The authors declare no conflict of interest.

References

1. De Olde, E.M.; Moller, H.; Marchand, F.; McDowell, R.W.; MacLeod, C.J.; Sautier, M.; Halloy, S.; Barber, A.; Benge, J.; Bockstaller, C.; et al. When experts disagree: The need to rethink indicator selection for assessing sustainability of agriculture. *Environ. Dev. Sustain.* **2017**, *19*, 1327–1342. [CrossRef]
2. Bockstaller, C.; Beauchet, S.; Manneville, V.; Amiaud, B.; Botreau, R. A tool to design fuzzy decision trees for sustainability assessment. *Environ. Model. Softw.* **2017**, *97*, 130–144. [CrossRef]
3. Rodias, E.C.; Lampridi, M.; Sopegno, A.; Berruto, R.; Banias, G.; Bochtis, D.D.; Busato, P. Optimal energy performance on allocating energy crops. *Biosyst. Eng.* **2019**, *181*, 11–27. [CrossRef]
4. Lampridi, M.G.; Sørensen, C.G.; Bochtis, D.D. Agricultural Sustainability: A Review of Concepts and Methods. *Sustainability* **2019**, *11*, 5120. [CrossRef]
5. Bockstaller, C.; Guichard, L.; Keichinger, O.; Girardin, P.; Galan, M.B.; Gaillard, G. Review article Comparison of methods to assess the sustainability of agricultural systems. A review. *Agronomy* **2009**, *29*, 223–235.
6. Snapp, S.S.; Grabowski, P.; Chikowo, R.; Smith, A.; Anders, E.; Sirrine, D.; Chimonyo, V.; Bekunda, M. Maize yield and profitability tradeoffs with social, human and environmental performance: Is sustainable intensification feasible? *Agric. Syst.* **2018**, *162*, 77–88. [CrossRef]
7. Allahyari, M.S.; Daghighi Masouleh, Z.; Koundinya, V. Implementing Minkowski fuzzy screening, entropy, and aggregation methods for selecting agricultural sustainability indicators. *Agroecol. Sustain. Food Syst.* **2016**, *40*, 277–294. [CrossRef]
8. De Olde, E.M.; Oudshoorn, F.W.; Sørensen, C.A.G.; Bokkers, E.A.M.; De Boer, I.J.M. Assessing sustainability at farm-level: Lessons learned from a comparison of tools in practice. *Ecol. Indic.* **2016**, *66*, 391–404. [CrossRef]
9. Sajjad, H.; Nasreen, I. Assessing farm-level agricultural sustainability using site-specific indicators and sustainable livelihood security index: Evidence from Vaishali district, India. *Community Dev.* **2016**, *47*, 602–619. [CrossRef]
10. Gaviglio, A.; Bertocchi, M.; Demartini, E. A Tool for the Sustainability Assessment of Farms: Selection, Adaptation and Use of Indicators for an Italian Case Study. *Resources* **2017**, *6*, 60. [CrossRef]
11. Peano, C.; Migliorini, P.; Sottile, F. A methodology for the sustainability assessment of agri-food systems. *Ecol. Soc.* **2014**, *19*, 19. [CrossRef]
12. Chopin, P.; Tirolien, J.; Blazy, J.M. Ex-ante sustainability assessment of cleaner banana production systems. *J. Clean. Prod.* **2016**, *139*, 15–24. [CrossRef]
13. De Luca, A.I.; Falcone, G.; Stillitano, T.; Iofrida, N.; Strano, A.; Gulisano, G. Evaluation of sustainable innovations in olive growing systems: A Life Cycle Sustainability Assessment case study in southern Italy. *J. Clean. Prod.* **2018**, *171*, 1187–1202. [CrossRef]
14. Rodias, E.; Berruto, R.; Bochtis, D.; Busato, P.; Sopegno, A. A computational tool for comparative energy cost analysis of multiple-crop production systems. *Energies* **2017**, *10*, 831. [CrossRef]
15. Bartzas, G.; Vamvuka, D.; Komnitsas, K. Comparative life cycle assessment of pistachio, almond and apple production. *Inf. Process. Agric.* **2017**, *4*, 188–198. [CrossRef]
16. Strapatsa, A.V.; Nanos, G.D.; Tsatsarelis, C.A. Energy flow for integrated apple production in Greece. *Agric. Ecosyst. Environ.* **2006**, *116*, 176–180. [CrossRef]
17. Rodias, E.; Berruto, R.; Bochtis, D.; Sopegno, A.; Busato, P. Green, yellow, and woody biomass supply-chain management: A review. *Energies* **2019**, *12*, 3020. [CrossRef]
18. Viola, I.; Marinelli, A. Life Cycle Assessment and Environmental Sustainability in the Food System. *Agric. Agric. Sci. Procedia* **2016**, *8*, 317–323. [CrossRef]
19. Mantoam, E.J.; Romanelli, T.L.; Gimenez, L.M.; Milan, M. Energy demand and greenhouse gases emissions in the life cycle of coffee harvesters. *Chem. Eng. Trans.* **2017**, *58*, 175.
20. Mantoam, E.J.; Romanelli, T.L.; Gimenez, L.M. Energy demand and greenhouse gases emissions in the life cycle of tractors. *Biosyst. Eng.* **2016**, *151*, 158–170. [CrossRef]
21. Sørensen, C.G.; Halberg, N.; Oudshoorn, F.W.; Petersen, B.M.; Dalgaard, R. Energy inputs and GHG emissions of tillage systems. *Biosyst. Eng.* **2014**, *120*, 2–14. [CrossRef]
22. Tassielli, G.; Renzulli, P.A.; Mousavi-Avval, S.H.; Notarnicola, B. Quantifying life cycle inventories of agricultural field operations by considering different operational parameters. *Int. J. Life Cycle Assess.* **2019**, *24*, 1075–1092. [CrossRef]

23. Bochtis, D.; Sorensen, C.G.; Kateris, D. *Operations Management in Agriculture*, 1st ed.; Elsevier: Amsterdam, The Netherlands, 2018; ISBN 9780128097168.
24. Edwards, W. Farm Machinery Selection. Available online: https://www.extension.iastate.edu/agdm/crops/html/a3-28.html (accessed on 3 January 2020).
25. Bochtis, D.D.; Sørensen, C.G.C.; Busato, P. Advances in agricultural machinery management: A review. *Biosyst. Eng.* **2014**, *126*, 69–81. [CrossRef]
26. Marinoudi, V.; Sørensen, C.G.; Pearson, S.; Bochtis, D. Robotics and labour in agriculture. A context consideration. *Biosyst. Eng.* **2019**, *184*, 111–121. [CrossRef]
27. Sørensen, C.G.; Nielsen, V. Operational analyses and model comparison of machinery systems for reduced tillage. *Biosyst. Eng.* **2005**, *92*, 143–155. [CrossRef]
28. Lampridi, M.G.; Kateris, D.; Vasileiadis, G.; Marinoudi, V.; Pearson, S.; Sørensen, C.G.; Balafoutis, A.; Bochtis, D. A Case-Based Economic Assessment of Robotics Employment in Precision Arable Farming. *Agronomy* **2019**, *9*, 175. [CrossRef]
29. Lee, J.; Cho, H.J.; Choi, B.; Sung, J.; Lee, S.; Shin, M. Life cycle assessment of tractors. *Int. J. Life Cycle Assess.* **2000**, *5*, 205–208. [CrossRef]
30. Aguilera, E.; Guzmán, G.I.; Infante-amate, J.; García-ruiz, R.; Herrera, A.; Villa, I. Embodied energy in agricultural inputs. Incorporating a historical perspective. *DT-SEHA 15*. 2015, p. 119. Available online: http://hdl.handle.net/10234/141278 (accessed on 5 December 2019).
31. Audsley, E.; Stacey, K.; Parsons, D.J.; Williams, A.G. Estimation of the greenhouse gas emissions from agricultural pesticide manufacture and use. *Uma ética para quantos?* 2014, p. 20. Available online: https://dspace.lib.cranfield.ac.uk/bitstream/handle/1826/3913/Estimation_of_the_greenhouse_gas_emissions_from_agricultural_pesticide_manufacture_and_use2009.pdf;jsessionid=DC1D51F03A8C73E065940B464D68BDBD?sequence=1 (accessed on 5 December 2019).
32. Mantoam, E.J.; Mekonnen, M.M.; Romanelli, T.L. Energy demand and water footprint study of an agricultural machinery industry. *Agric. Eng. Int. CIGR J.* **2018**, *20*, 132.
33. Kitani, O. *CIGR Handbook of Agricultural Engineering, Volume 5: Energy and Biomass Engineering*; American Society of Agricultural and Biological Engineers: St. Joseph, MI, USA, 1999.
34. ASABE. *ASAE D497.5 FEB 2006 Agricultural Machinery Management Data*; ASABE: St. Joseph, MI, USA, 2006.
35. Kitani, O.; Jungbluth, T.; Peart, R.; Ramdani, A. *CIGR Handbook of Agricultural Engineering Volume V*; CIGR: Liege, Belgium, 1999; ISBN 0929355970.
36. Canakci, M.; Akinci, I. Energy use pattern analyses of greenhouse vegetable production. *Energy* **2006**, *31*, 1243–1256. [CrossRef]
37. Kuswardhani, N.; Soni, P.; Shivakoti, G.P. Comparative energy input-output and financial analyses of greenhouse and open field vegetables production in West Java, Indonesia. *Energy* **2013**, *53*, 83–92. [CrossRef]
38. Reineke, H.; Stockfisch, N.; Märländer, B. Analysing the energy balances of sugar beet cultivation in commercial farms in Germany. *Eur. J. Agron.* **2013**, *45*, 27–38. [CrossRef]
39. Schramski, J.R.; Jacobsen, K.L.; Smith, T.W.; Williams, M.A.; Thompson, T.M. Energy as a potential systems-level indicator of sustainability in organic agriculture: Case study model of a diversified, organic vegetable production system. *Ecol. Model.* **2013**, *267*, 102–114. [CrossRef]
40. Busato, P.; Berruto, R. Minimising manpower in rice harvesting and transportation operations. *Biosyst. Eng.* **2016**, *151*, 435–445. [CrossRef]
41. ASABE. *D497.7: Agricultural Machinery Management Proposed*; ASABE: St. Joseph, MI, USA, 2015.
42. ASAE. *ASAE EP496.3—Agricultural Machinery Management*; ASABE: St. Joseph, MI, USA, 2009.
43. Tsatsarelis, C. *Agricultural Machinery Management*, 1st ed.; Giachoudi Publications: Thessaloniki, Greece, 2006; ISBN 960-7425-86-3.
44. Saunders, C.; Barber, A.; Taylor, G. *Food Miles-Comparative Energy/Emissions Performance of New Zealand's Agriculture Industry*; Agribusiness and Economics Research Unit, Lincoln University: Canterbury, New Zealand, 2006.

energies

MDPI

Article

Decision Support System to Implement Units of Alternative Biowaste Treatment for Producing Bioenergy and Boosting Local Bioeconomy

Christos Vlachokostas [1,*], Charisios Achillas [2], Ioannis Agnantiaris [1], Alexandra V. Michailidou [1], Christos Pallas [3], Eleni Feleki [1] and Nicolas Moussiopoulos [1]

1 Laboratory of Heat Transfer and Environmental Engineering, Department of Mechanical Engineering, Aristotle University Thessaloniki, P.O. Box 483, 54124 Thessaloniki, Greece; j.agnadiaris@gmail.com (I.A.); amicha@meng.auth.gr (A.V.M.); efeleki@meng.auth.gr (E.F.); moussio@eng.auth.gr (N.M.)

2 Department of Supply Chain Management, International Hellenic University, Kanelopoulou 2, 60100 Katerini, Greece; c.achillas@ihu.edu.gr

3 Municipality of Serres, Merarchias Avenue 1, 62122 Serres, Greece; palas@serres.gr

* Correspondence: vlahoco@auth.gr

Received: 11 April 2020; Accepted: 30 April 2020; Published: 6 May 2020

Abstract: Lately, the model of circular economy has gained worldwide interest. Within its concept, waste is viewed as a beneficial resource that needs to be re-introduced in the supply chains, which also requires the use of raw materials, energy, and water to be minimized. Undeniably, a strong link exists between the bioeconomy, circular economy, bioproducts, and bioenergy. In this light, in order to promote a circular economy, a range of alternative options and technologies for biowaste exploitation are currently available. In this paper, we propose a generic methodological scheme for the development of small, medium, or large-scale units of alternative biowaste treatment, with an emphasis on the production of bioenergy and other bioproducts. With the use of multi-criteria decision analysis, the model simultaneously considers environmental, economic, and social criteria to support robust decision-making. In order to validate the methodology, the latter was demonstrated in a real world case study for the development of a facility in the region of Serres, Greece. Based on the proposed methodological scheme, the optimal location of the facility was selected, based on its excellent assessment in criteria related to environmental performance, financial considerations, and local acceptance. Moreover, anaerobic digestion of agricultural residues, together with farming and livestock wastes, was recommended in order to produce bioenergy and bioproducts.

Keywords: bioenergy; efficiency of bio-resources; decision support system; multi-criteria analysis; sustainability

1. Introduction

Circular economy and resource efficiency have received increasing attention in research and environmental policy agenda in recent years. Circularity can be a catalyst for productive reconstruction and has a clear regional dimension, where the value of waste is a key element. Within the concept of circular economy, waste, energy, and water consumption need to be minimized [1–4] and, in any case, the corresponding environmental pressures should respect target or limit values [5]. Bioeconomy focuses on the use of biomass/biowaste in primary production procedures (e.g., agriculture, forestry, fisheries) and substantial valorization of raw materials. Biomass is expected to substantially support the achievement of the EU renewable energy targets [6].

Undoubtedly, there is a strong linkage between the bioeconomy, circular economy, bioproducts, and bioenergy. Streams of excess or end-of-life materials, previously regarded as waste, that originate from anthropogenic activities can be channeled through available technologies (e.g., anaerobic digestion, gasification, pyrolysis) and transmute into useable energy carriers, organic biofertilizer abundant in nutrients, and, in general, original and innovative materials. A characteristic example is the production of biofuel (biogas or syngas), which can be characterized as a major means of energy recovery from biowaste streams. An appreciable amplitude of different technological solutions is currently available for small, medium, or large-scale units of alternative biowaste treatment (UABT), with an emphasis on the production of bioenergy and other bioproducts. Such solutions are based on applications that extend from tailor-made systems, which incorporate existing available facilities (e.g., pumps and storages in an existing farm, buildings for the combined heat and power (CHP) installation etc.), to specialized concepts that are diverse, where the main parts have been pre-manufactured.

Although there are many solutions, biowaste disposal in landfills, or, in the worst case, in uncontrolled open dumps, is still an existing practice internationally [7]. This cannot be considered as a proper managerial approach for biowaste, having in mind the intense environmental load imposed. Apart from aesthetic degradation, impacts comprise air pollution and Green House Gas (GHG) emissions, soil and water contamination, reduced land values, and landscape blight. Consequently, optimal biowaste management constitutes a critical activity that reinforces environmental sustainability by minimizing impacts, especially related to uncontrolled discarding [8–12].

The appropriateness and benefits of each technological solution should be examined carefully to support treatment decisions, considering the type of biowaste, local characteristics, and conditions [13–15]. In the material to follow, a generic methodological scheme was developed and demonstrated to support optimal decision-making in promoting and implementing UABTs. The basic structure and components combine environmental, economic, and social parameters and multi-criteria decision analysis (MCDA) for the optimal site selection of a UABT. Since the early 1980s, Ross and Soland [16] reported that problems involving the location and promotion of such units include multi-criteria considerations and ought to be modelled likewise. Nevertheless, the literature has scarcely studied special waste streams, such as biowaste, centralizing mostly on municipal solid waste (MSW) management [17–22]. MCDA has gained wide acceptance in recent years over quantitative modeling, as MCDA embodies both quantitative as well as qualitative variables [23] and can be tractably combined with other tools, such as life cycle assessment, ecological footprint, and environmental indicators [24]. The approach includes the social criterion (e.g., NIMBY - Not In My Back Yard syndrome), which is usually disregarded [25,26]. It also assists relevant stakeholders, public authorities, and producers by providing a roadmap to the feasibility and essential steps for the development of a micro-to-medium-scale localized UABT. Localization is a prerequisite to achieve economies of scale and an emphasis is given to an optimal location, and the production of bioenergy and other bioproducts [27,28].

In this context, the present work frames a consistent generic methodology and seeks optimal UABT solutions, based on multiple criteria and parameters. More specifically, apart from defining the optimal site for the installation of a new UABT, the proposed methodology also assesses the recommended technological solution, based on the characteristics and quantities of locally available biowaste, as well as its size and capacity. The methodology is demonstrated for a real-world case in the region of Serres, Greece.

2. Methodology: Basic Structure and Components

2.1. Selection of Area and Inventory Analysis

Figure 1 presents the main components and the basic steps of the methodological scheme. More specifically, the first step is to determine the wider area for the development of a UABT (Step 1). Crucial aspects for this decision are the availability, quantity, and quality of biowaste regionally.

Characteristic typologies can be found in the Waste Framework Directive [29] and related regulations, e.g., park and yard waste, animal waste (feces, urine, and manure), kitchen and food waste, vegetal waste (food preparation and products), household and similar wastes, common sludges, mixed waste, undifferentiated materials, etc. To assure smooth replenishment and long-term sustainability of a UABT, it is important to pre-assess the continuing availability and seasonality of biowaste locally. Regional logistics infrastructure, mainly transportation and storage facilities, also need to be considered for the available feedstock, due to their criticality in the unit's viability, since the transportation cost is a crucial parameter for the viability of the logistics network [30]. All the above data can be the output of a strengths, weaknesses, opportunities, and threats analysis, for the area under consideration. This initial stage of the methodology puts forward a pool of available types of biomass as potential feedstock for an UABT.

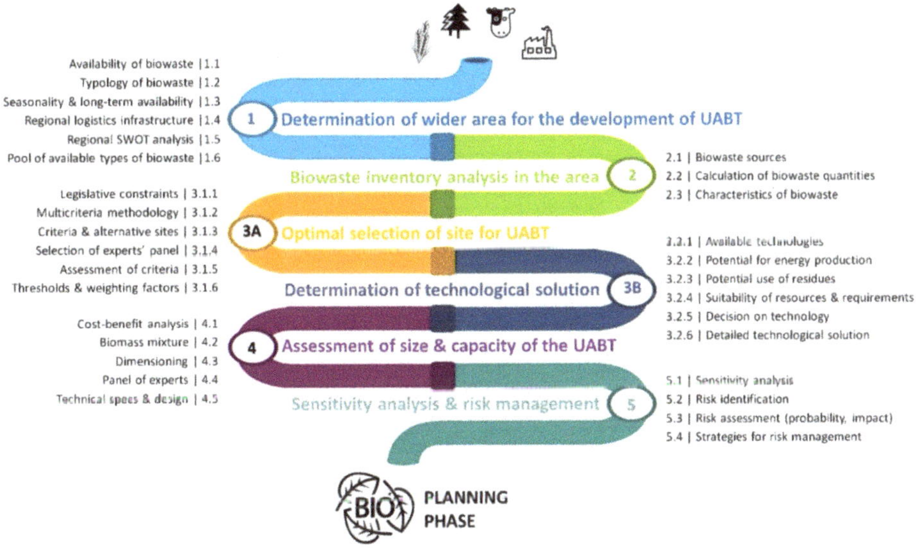

Figure 1. Basic structure and components of the methodological framework.

In the case that an area is considered "eligible" for the development of a UABT, the second step is the preparation of a detailed biowaste inventory analysis that (i) identifies meticulously the biowaste sources, (ii) calculates the available biowaste quantities, and (iii) describes the characteristics of the available biowaste (Step 2). Such an inventory will detail the availability of exploitable biowaste within the region under study and will assist decision-making.

2.2. Optimal Selection of the Site

The inventory analysis is followed by the multiple criteria decision analysis (MCDA) for the optimal location of the UABT (Step 3a). This process is strongly interrelated and usually realized in parallel to the selection of the most appropriate technology (Step 3b). Multi-criteria mathematical modeling examines criteria that are usually in conflict in the decision-making process [31–36]. On this basis, and after screening all legislative constraints (e.g., environmental permits), the appropriate multi-criteria analysis technique needs to be determined. The literature reports a considerable number of techniques available, with different characteristics and uses [37,38]. Multi-criteria analysis techniques are more or less suitable depending on the special characteristics of the case under consideration [39]. In the scientific literature, modelling of waste management solutions mostly takes into account the economic and environmental dimensions of the problem, whilst the social concerns are usually neglected.

Morrissey and Browne [26] meticulously reported sustainable waste management applications. It should be noted that the social pillar constitutes a crucial matter for the future viability of the investment in cases like the one examined in the material to follow. On this basis, social concerns should be simultaneously considered for deciding on the optimal UABT site.

The methodology developed herein adopts the ELECTRE III technique [40]. A main characteristic of this technique is the use of the pseudo-criteria concept in order to depict the different angles of the studied case/problem. ELECTRE III utilizes three pseudo-criteria and initiates a robust comparison of each option with another in relation to all criteria included in the analysis. An ascending and descending distillation process forms the basis for the construction of two complete pre-orders. The intersection of the two complete pre-orders produces a final classification of the alternatives. An advantage of this technique is the sensitive analysis capability. In the framework of the sensitivity analysis, scenarios for the values of the main parameters are performed in order to further test the solution by observing the effect of the adopted values' variation to the result. Comparative assessment of the pre-orders for each scenario leads to a final robust outcome or to a model re-analysis [41]. This technique is widely used in the examination of environmental problems and waste management issues, which provides an advantage to the others [20,33,42–44].

The designation of three thresholds is a prerequisite, i.e., (i) preference threshold (p), (ii) indifference threshold (q), and (iii) veto threshold (v). ELECTRE III uses these thresholds in its mathematical rationale in order to better include real-life uncertainties [41]. Another advantage of this multi-criteria technique is the simultaneous consideration of quantitative (e.g., price of sites, distances, etc.) and qualitative criteria (e.g., landscape degradation, aesthetics, social acceptance, etc.), as this approach depicts a good fit of the data in such applications.

It goes without saying that a meticulous investigation in the wider area under study is needed in order to landmark all the available locations appropriate for the installation of a UABT. As a next step, the assessment of different, usually conflicting, criteria is performed in parallel with defining the weighting factor of each criterion. The weighting factor expresses its relative significance in comparison to the others. A clear definition of the parameters included in the analysis is necessary in order to reliably value all available alternatives and perform the pairwise comparison process. After defining the criteria and weighting factors, the related information and data should be assembled. Multi-criteria evaluation of sites for the UABT location is mathematically formulated with the use of a set of criteria ($Cr_1, Cr_2, Cr_3 \ldots$) applicable to a set of alternatives ($A_1, A_2, A_3 \ldots$). $V_j(A)$ mathematically expresses the evaluation of alternative A i, for the criterion j. ELECTRE III is a ranking method, thus it puts forward a ranking prioritization, which is grounded on binary outranking relations for two interrelated concepts, i.e., (i) concordance (c_j) and (ii) non-discordance (d_j). More specifically, the concept of the concordance relation is applicable when alternative A_1 outranks alternative A_2 in case a sufficient majority of criteria are in favor of alternative A_1. The non-discordance relation is applicable when the concordance condition holds, and none of the criteria in the minority should be opposed strongly to the outranking of A_2 by A_1. The credibility index characterizes the assertion that A_1 outranks A_2 and illustrates the true degree of this assertion [45]. A pair of alternatives is compared for each criterion with the use of pseudo-criteria, namely the indifference (q_j) and preference (p_j) for which they apply:

- When $V_j(A_1) - V_j(A_2) \leq q_j$, then no difference between A_1 and A_2 is identified for the specific criterion j, thus $c_j(A_1, A_2) = 0$.
- When $V_j(A_1) - V_j(A_2) > p_j$, then A_1 is strictly preferred to A_2 for criterion j, thus $c_j(A_1, A_2) = 1$.

The concordance index $c_j(A_1, A_2)$ of each criterion j is mathematically formulated as follows:

$$\begin{array}{l} V_j(A_1) - V_j(A_2) \leq q_j \Leftrightarrow c_j(A_1, A_2) = 0 \\ q_j < V_j(A_1) - V_j(A_2) < p_j \Leftrightarrow c_j(A_1, A_2) = \frac{V_j(A_1) - V_j(A_2) - q_j}{p_j - q_j} \\ V(A_1) - V(A_2) \geq p_j \Leftrightarrow c_j(A_1, A_2) = 1 \end{array}.$$

A global concordance index $C_{A_1A_2}$ for each pair (A_1, A_2) is calculated with the use of c_j (A_1, A_2) as mathematically illustrated below:

$$C_{A_1A_2} = \frac{\sum_{j=1}^{n} w_j \cdot c_j(A_1, A_2)}{\sum_{j=1}^{n} w_j}, w_j \text{ expresses the weighting factor of criterion j.}$$

The discordance index (d_j) is computed with the adoption of indifference (q_j), preference (p_j), and the veto threshold (v_j), which expresses the maximum acceptable difference for not rejecting the assertion A_1 outranks A_2. More specifically:

- When V_j (A_1) − V_j (A_2) ≤ p_j, no discordance exists and therefore d_j (A_1, A_2) = 0.
- When V_j (A_1) − V_j (A_2) > v_j, then d_j (A_1, A_2) = 1.

Thus, d_j (A_1,A_2) is mathematically formulated as:

$$\begin{cases} V_j(A_2) - V_j(A_1) \le p_j \Leftrightarrow d_j(A_1, A_2) = 0 \\ p_j < V_j(A_2) - V_j(A_1) < v_j \Leftrightarrow d_j(A_1, A_2) = \frac{V_j(A_2) - V_j(A_1) - p_j}{v_j - p_j} \\ V_j(A_2) - V_j(A_1) \ge v_j \Leftrightarrow d_j(A_1, A_2) = 1 \end{cases}.$$

The index of credibility $\delta_{A_1A_2}$ of the claim A_1 outranks A_2 is then formulated as:

$$\delta_{A_1A_2} = C_{A_1A_2} \prod_{j \in \bar{F}} \frac{1 - d_j(A_1, A_2)}{1 - C_{A_1A_2}}, \text{ with } \bar{F} = \{j \in F, d_j(A_1, A_2) > C_{A_1A_2}\}.$$

The claim A_1 outranks A_2 is rejected if v_j is exceeded for at least one of the criteria under consideration. After constructing the ranking scheme, as a last step, sensitivity analysis is activated, considering that values and assessments are often subjective in real-world cases and originate from less or more reliable estimations (criteria qualitative expression, threshold parameters, weighting factors, etc.). As already highlighted, this is considered a major advantage of the adopted MCDA technique due to the high data uncertainty in the thematic area under study [46].

2.3. *Technology, Size, and Capacity of the UABT*

The decision for the optimal site in the wider area presents a strong interrelationship with the determination of the technology to be adopted (Step 3b). A range of available technologies are available (e.g., anaerobic digestion, aerobic digestion, gasification, pyrolysis) and possibly a combination of these. The key factors that need to be considered for selecting the optimal technology are the potential for bioenergy production (electricity, heat or gas), possible utilization of the residues of each technological solution (e.g., biofertilizer), suitability of available bioresources, and pre-treatment requirements. All this information supports the optimal decision on technology for the UABT, based on constraints related to the type of biowaste and specific technological solutions used.

As a next step, the assessment of the size and capacity of the UABT is realized (Step 4), which is based on a cost-benefit analysis (CBA) and the determination of the biomass mixture for the UABT (range of sources depending on the biowaste quality). The latter ultimately leads to the dimensioning and detail of the technical specifications for the UABT. Highlighting strategies for risk management is the last step of the methodological roadmap presented (Step 5). The abovementioned methodological framework is expected to support policy-makers to facilitate the conceptual phase towards the development of a UABT, before advancing to the next phase of the planning phase where more resources are required. In this light, decision-makers may screen available options and give a green light for further analysis only to those cases where the endeavor is proven viable.

3. Application of Methodology in the Region of Serres, Greece

3.1. The Area under Study

The applicability of the methodological framework was validated in a real-life case study in the region of Serres, Greece. The region of Serres under study was pre-selected based on the considerable regional agricultural activity and biomass availability. The population of the wider Serres area is approximately 200,000, out of which 25% is urban, 20% semi-urban, and 55% rural/agricultural. Its economic activity mainly deals with the primary sector and especially with agriculture (approximately 60%). In the region, there are 162,800 ha of cultivated land and 113,300 ha of pasture land. More than 90% of the cultivated land is private. In the area, there is well-developed agricultural production, which contributes 3.4% of the total national agricultural product, with the main drivers being dairy and meat production. Apart from farmers, within a range of 20 km from the municipality of Serres, there are a number of agri-food manufacturers. Specifically, the relevant production activity consists of 2 leading dairy industries, 6 small cheese/yogurt production units, 2 olive mills, 1 potato packing and processing unit, 4 poultry farms, more than 30 cattle and pig rearing units for dairy and meat production, and one slaughterhouse.

These facilities are capable of supplying adequate quantities of biowaste to generate a mixture that ensures the continuous operation of a UABT. Considering the available data and given the population and basic economic activity characteristics of the wider area under study, a primary course of action regarding biowaste energy content could be supported by the recent legislative framework set in Greece regarding energy communities [47]. The legislation sets the rules for the incorporation of quasi-private entities with the option of participation of the municipalities as shareholders, while also providing the alternative of subsidized (or with tax-incentives) renewable energy plants. Such strategic schemes could also opt from the formation of public–private partnerships.

The region under study has great potential for the application of a productive UABT model due to many reasons, such as: (i) The availability of resources and biowaste, (ii) scientific potential and know-how, and (iii) the existence of a primary sector with growth potential and needs for modernization and reduction of production costs. Last, but not least, local farmers are strongly interested in adjusting their traditional economic attitudes. This high level of engagement was is result of intense awareness-raising activities (including flyers/brochures dissemination, press and media publicity, public presentations, visits to demonstration projects) as well as public opinion surveys that leaded to accurate expectation mapping of all involved local stakeholders [48].

3.2. Alternative Sites and MCDA

According to the methodology described, different locations for the development of the UABT within the area under study (Figure 1) were promoted. Following a detailed survey on land availability, four alternative available fields were selected, such as those sites are summarized in Table 1.

Next, the criteria for the analysis were defined in order to interlace financial viability, environmental performance, and acceptance by the local community, taking additionally into account all legislative constraints. A critical mass of local experts, also aware of the local special characteristics, were invited. In total, 13 experts representing local and regional authorities and the business sector in the study area were interviewed to decide on the criteria to be used in the case under study. This resulted in the following criteria: (Cr_1) Distance from biowaste source, (Cr_2) accessibility of the site with the existing infrastructure, (Cr_3) distance from units for the consumption of energy (electricity and heat) produced, (Cr_4) distance from farms for compost use, (Cr_5) value of land, (Cr_6) impacts on local populations (e.g., air pollution, odors, noise pollution), (Cr_7) impacts on local ecosystems, (Cr_8) aesthetic degradation, (Cr_9) increase in road traffic due to UATB's operation, and (Cr_{10}) social acceptance.

Table 1. Alternative locations for UABT development in the area under study.

Location of Sites

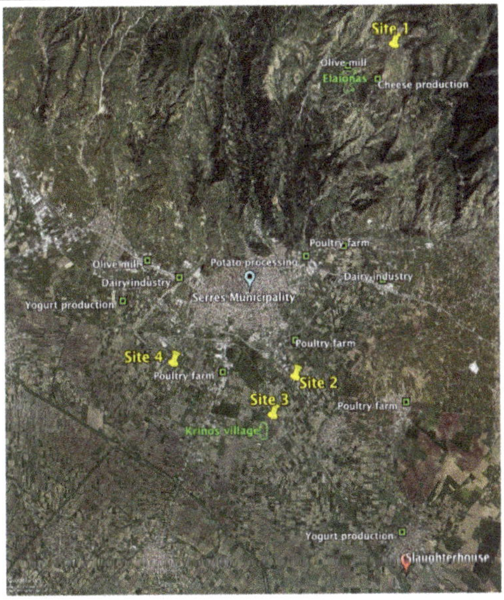

Site	Description
Site 1	Located in a public area in Eleonas in the Forage Park. Access is through a forest road network with great difficulty. Road and water supply projects are required. Distance from the city is about 10 km and 23 km from the Slaughterhouse.
Site 2	Located in a public area in the Farm of Serres, 2 km from Serres and 7 km from the Slaughterhouse. Access is easy via the "Serres-Neos Skopos" provincial road. A few meters from the site, there is a municipal vegetable garden and a greenhouse.
Site 3	Located in a municipal area in the Farm of Serres, 0.5 km from Krinos village and 8 km from the Slaughterhouse (via dust road, or 12 km via municipal roads).
Site 4	Located in a public area in the Farm of Serres, near the Omonia Sports Park. The site is 13.5 km from the Slaughterhouse via municipal roads and 0.5 km from the city.

The four site locations were assessed over their performances on the 10 designated criteria (Table 2). Criteria Cr_2, Cr_6, Cr_7, Cr_8, Cr_9, and Cr_{10} were qualitatively assessed by the experts involved in the survey. Those criteria that are related to the distances (Cr_1, Cr_3, Cr_4) were assessed with the use of Google maps and the location of available biowaste sources and facilities where the outputs of the UABT could be exploited (farms and manufacturers) [48]. Cr_5, which is related to the value of the land, was assessed with mean objective values, which were provided by the Hellenic Statistical Authority. For all selected criteria, the higher the performance, the more preferable the alternative is assessed.

Table 2. Assessment of alternative site locations (*in 1–10 scale, where 10 is the most preferable*).

Criterion	Site 1 (1–10)	Site 2 (1–10)	Site 3 (1–10)	Site 4 (1–10)	Weights (%)	p_i	q_i
Cr_1 - Distance from biowaste source	1.46	8.92	6.92	5.69	15	1.87	0.56
Cr_2 - Accessibility of site	1.69	9.62	6.31	6.77	14	1.98	0.59
Cr_3 - Distance from units for energy consumption	2.00	9.46	7.54	7.15	12	1.87	0.56
Cr_4 - Distance from farms for compost use	3.85	9.23	8.31	7.69	8	1.35	0.40
Cr_5 - Value of land	2.85	9.15	7.62	7.23	12	1.58	0.47
Cr_6 - Impacts on local populations	9.23	7.31	2.62	2.00	10	1.81	0.54
Cr_7 - Impacts on local ecosystems	3.85	8.00	5.92	5.77	10	1.04	0.31
Cr_8 - Aesthetic degradation	4.85	8.46	4.69	3.23	6	1.31	0.39
Cr_9 - Increase in road traffic due to UATB	3.08	7.08	2.85	3.46	5	1.06	0.32
Cr_{10} - Social acceptance	8.69	8.46	2.69	2.23	8	1.62	0.48

Weighting factors are also depicted in Table 2. Those were estimated as averages of the experts' views. In order to overcome subjectivity issues, sensitivity analysis was employed. In our case, a low computational time was required to re-calculate the optimal solutions with the modified parameters. The calculation of p_i (preference thresholds) for the selected criteria was based on the use of Equation (1) [44,46,49,50], while the calculation of q_i (indifference thresholds) was based on the use of Equation (2) [46,51]:

$$p_i = \frac{1}{n}(V_i max - V_i min), \quad (1)$$

$$q_i = 0.3 \cdot p_i \quad (2)$$

3.3. Technology, Size, and Capacity of the UABT

Based on the above analysis, a combination of anaerobic and aerobic digestion is promoted as the selected technological solution. Anaerobic digestion (AD) fits the local agricultural and farming well since manure, cheese whey, side products (i.e., rotten potato pulp, oil mill waste), and slaughterhouse waste are efficient substrates available in the area. AD is considered as the process where value is produced mainly for the generation of bioenergy (CHP) and biofertilizer. AD is a fermentation process, which is realized in an air-sealed biodigester (under conditions of oxygen absence). The feedstock (biowaste substrates) is converted into fuel gas and digestate (as a bio-product). The biogas that is produced comprises mainly of CH_4 (50%–70%) and CO_2 (30%–50%). Moreover, in smaller quantities, the biogas consists of H_2O vapor, H_2S, and other elements. The digestate is produced from the digestion of substrates, after biogas extraction. Wet digestate comprises of nutrients (e.g., nitrogen, lignin, phosphorous), inorganic salts (e.g., ammonium, phosphate, potassium), as well as other minerals. Therefore, it is suitable for use as a naturally derived fertilizer. In the literature, there are many studies related to the exploitation of AD [52–55]. Regional AD applications are also detailed in [56]. The substrates used determine the appropriate technical infrastructure that is required (e.g., pipes' sizing, pumps' specifications, gas storage requirements, gas treatment technology, and Combined Heat and Power - CHP design).

The technology adopted does not add to the CO_2 load in the atmosphere, since the CO_2 produced is offset by the prevented emissions of CH_4 in the case that slurry is stored in open spaces. In this light, biogas can significantly contribute to decarbonization of the economy. Furthermore, decentralized bioenergy generation from biofuel in UABTs with CHP is becoming more attractive for the cases where generated heat is exploited in facilities within the proximity of the UABT. Moreover, the use of UABTs' end-products (energy and bioproducts) needs to be meticulously examined within the pre-planning phase. In this context, the use of the produced bio-fertilizer in farms close to the developed UABTs highly supports their financial performance and effectiveness. The sufficiency of arable land for spreading available biofertilizer and an available adjacent market for the digestate needs to be investigated. In addition, biofertilizers' nutrients are predominately contained in a mineral form, which allows ease of absorption by plants and increased uptake efficiency, in comparison to nutrients in raw manure or in slurry [57].

Based on the above, it is proven in practice by the operation of such multi-functional facilities that this technology offers reduced local community energy costs, low-cost and environmentally safe recycling of manure and biowaste, cheap and efficient crop biofertilization means, and a reduction of odors due to intensive farming activity. Thus, considering also that in many areas, organic substances are still disposed in landfills (contributing to local CH_4 emissions), the proposed UABT apparently dually contributes towards global warming mitigation through (a) the decreased CH_4 emissions in animal farms (due to digestate management and avoidance of open slurry storage), and (b) the production of a green decarbonized (bio-)fuel. Along with the latter, improved digestate nutrient management, in combination with sustainable agri-practices [58], has a positive effect on the reduction of NH_3 and NO_x emissions, as well as the eutrophication of surface and ground water since leakages can be decreased or even avoided. Furthermore, local eutrophication is also achieved through biogas treatment of manure [59].

4. Results and Discussion

4.1. Feedstock

Based on the inventory analysis, the UABT receives organic biowaste types. More specifically, Table 3 presents the results of the annual received quantity of each substrate, their supply (d/a), and the quantity of each substrate per day of supply [48]. The codification of each substrate is presented in accordance with the European Waste List (EWL) (as per Annex to Decision 2000/532/EC).

Table 3. Substrate mix composition, yearly supply schedule, and yearly supply quantities.

Raw Material	Supply Quantity (Tons/Day)	Days of Supply (Days/Year)	Overall Quantity (Tons/Year)	EWL Code
Dairy cow slurry manure	6.27	200	1254	02 01 06
Cow manure	5.91	200	1183	02 01 06
Calf manure	0.51	200	102	02 01 06
Cattle manure	0.73	200	146	02 01 06
Poultry manure	0.32	104	33	02 01 06
Cheese whey	1.78	156	278	02 05 01
Rotten potato pulp	1.00	32	32	02 01 03
Olive mill waste	2.37	32	76	02 03 01
Slaughterhouse waste - Intestine content	1.44	104	150	02 02 01
Slaughterhouse waste - Stomachs and fat	0.25	52	13	02 02 02
Slaughterhouse waste - Blood	0.35	52	18	02 02 02
		TOTAL	**3285**	

The energy density of manure is considerably low, mainly due to the high content in water, along with the relatively low gas yield. The latter decreases its attractiveness for the case of long transportation distances. On the other hand, hydraulically, slurry is comparatively easy to handle. In the cases where the proportion of manure is relatively high within the digester, substrates like grass or solid manure, which are usually hydraulically more demanding, show a good performance in a small-sized biogas plant.

4.2. Optimal Site

The use of ELECTRE III in the case herein examined was realized with LAMSADE software. The two distillations (ascending and descending) that result from the analysis of the sites' performances are graphically displayed in Figure 2a. Both ascending and descending distillations promote Site 2 as the optimal alternative for the UABT's location. This is also displayed in Figure 2b, where the four alternatives ae hierarchically ranked. Site 2 (optimal location) is approximately 1.5 km south-southwest of the city of Serres, within municipality-owned land of approximately 16 acres.

As presented in Figure 2b, site 2 is judged as the optimal location locally. The latter is grounded on its outstanding assessment in most of the examined criteria (Table 2). This is well-justified due to the fact that site 2 is in the proximity of major producers of high-value biomass and potential end-users of the produced energy. Due to the fact that the area is industrial, the infrastructure in the proximity of site 2 is in very good condition, while the specific site also shows a very good performance in the social criteria (impacts on local populations and ecosystems, aesthetic degradation, increase in road traffic, social acceptance).

As a next step, a sensitivity analysis was conducted. The thresholds in the mathematical formulation (preference and indifference) were modified, as presented in Table 4. More specifically, six scenarios were examined. The optimal location for the development of the UABT in the region of Serres is site 2 for all examined scenarios. The latter increases the robustness of the basic solution and provides confidence in relation to the efficiency of the location over other available alternatives.

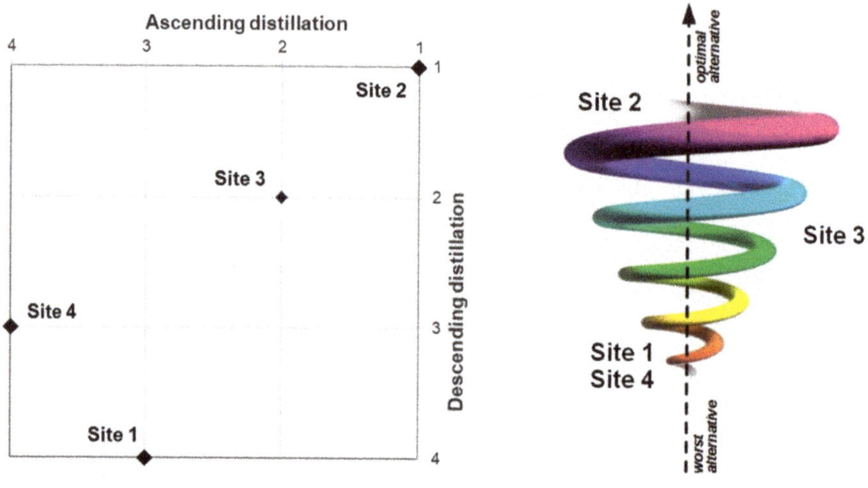

(**a**) (**b**)

Figure 2. Optimal UABT location: (**a**) Ascending and descending distillations - (**b**) Ranking of alternative locations.

Table 4. Sensitivity analysis.

Thresholds	**Scenario**						
	Baseline	A	B	C	D	E	F
p_i	p_i	$1.25 \times p_i$	$1.50 \times p_i$	$0.75 \times p_i$	$0.50 \times p_i$	p_i	$0.50 \times p_i$
q_i	$0.3 \times p_i$	$0.375 \times p_i$	$0.45 \times p_i$	$0.225 \times p_i$	$0.15 \times p_i$	0	$0.30 \times p_i$
Variation	-	+25%	+50%	−25%	−50%	$p_i = p_i \mid q_i = 0$	$p_i = \frac{1}{2} p_i \mid q_i = 0$
Ranking	**1st: Site 2** 2nd: Site 3 3rd: Site 1,4	**1st: Site 2** 2nd: Site 3,4 4th: Site 1	**1st: Site 2** 2nd: Site 3,4 4th: Site 1	**1st: Site 2** 2nd: Site 3 3rd: Site 1,4	**1st: Site 2** 2nd: Site 3 3rd: Site 1,4	**1st: Site 2** 2nd: Site 3 3rd: Site 1,4	**1st: Site 2** 2nd: Site 3 3rd: Site 1,4

4.3. Technical Specifications

The biowaste is collected from the production and disposal sites and transported by container trucks and/or simple trucks to the UABT. Biowaste is introduced into the reception tanks to be mixed through agitation and formulate the final substrates' mix for AD. The mix is pumped and transferred (via the displacement pumps) to the digesters. The UABT will produce approximately 160,000 m^3 of biofuel per year. The biogas is collected and transferred to the internal combustion engine (ICE) for the production of electrical and thermal energy. The estimated electricity generation amounts to 311,996 kWh/a of operation, including self-consumption needs and grid losses. The estimated thermal bioenergy generation is 356,216 kWh/a, including thermal losses and self-thermal energy consumption. More specifically, the excess electrical energy is fed to the low voltage grid and its selling price offsets the electricity purchase costs of the municipality of Serres. The excess thermal energy serves the heating needs of a greenhouse adjacent to the unit's site and is partially used to heat the substrates' mix inside the biodigesters through a shell-type heat exchanger. This conducts the heat from the ICE's exhaust gases to the closed hot water circuit passing through the digesters. At the closed hot water circuit's return, the water, at a lower temperature than that when coming out of the shell-type heat exchanger, is pre-heated with the use of a heat exchanger (plate-type) and simultaneously cools the ICE, before passing through the shell-type heat exchanger again and increasing its temperature from the exhaust gases of the ICE.

The above process was calculated to produce 3121 t of aqueous residue, rich in nutrients for deposition in cultivations adjacent to the UABT. The proposed AD process is thermophilic. The biowaste mixture remains inside the digesters for at least 20 days (hydraulic retention time ≥ 20 days) at a temperature of at least 52 °C. The digestate is pumped in a controlled manner and transported to the screw separator, which recovers its solid fraction to be bagged, while the liquid fraction is placed in a lagoon. The solid fraction bags are picked up by trucks and the liquid fraction is pumped and transported by tank containers, both to be deposited in adjacent crops during eligible fertilization periods.

It should be emphasized that the application of biofertilizer needs to be realized in the growth season for any given crop. This is necessary in order to increase the uptake of nutrients, as well as to avoid a leaching or surplus of nutrients. Apart from the nutrient content, a fertilizer needs to also show high quality in sanitation and the reduction of pathogens; quantities of organic compounds and heavy metals, plastic, or other contaminants; etc.

The proposed methodology's clear strength lies on the multi-functionality of this technological solution. UABTs' sustainability can be grounded on many different aspects, such as (a) exploitation of organic waste, (b) upgraded management of manure, (c) increased uptake efficiency of nutrients, (d) reduced odors, (e) protection of the environment and reduced GHG emissions, (f) increase of material value, and (g) bioenergy (in the form of electricity and heat) and biofuel production.

5. Conclusions

In our work, we proposed a methodological scheme aimed towards supporting decision-makers in assessing the development of a UABT, based also on multiple criteria and factors. The methodology was effectively validated in a real-world case study in the region of Serres, Greece. In the approach proposed, the decision for the optimal location of the UABT within the area under study was based on the results of a multicriteria analysis (with the use of ELECTRE III technique), taking into consideration qualitative and quantitative criteria. In the case presented, the optimal site for the location of the UABT results from its excellent performance in the selected criteria, which represent all pillars of sustainable development, namely financial viability, public acceptability, and environmental protection.

The proposed methodological scheme provides a roadmap either for public bodies or private companies that aim to invest in green energy projects. The methodology followed in this work is generic so as to be followed under different requirements and constraints in any real such investment, with slight modifications of the criteria or values in the thresholds and weighting factors according to the special conditions in the area under study. The methodological framework can also be used to make decisions on other issues related to the supply chain management, namely the optimal location of collection sites, warehousing, sorting centers, etc., after relevant modifications in the criteria considered.

Even though the production of biogas has been widely realized over the past decades, the large-scale substitution of fossil fuels is still considered a major innovation in the energy sector. Energy technologies that exploit biomass for energy production are still non-competitive and technologically inefficient and immature against the conventional use of non-renewable sources. The methodology presented provides a little stepping stone towards supporting a bioeconomy within the field of clean technology biowaste and residues, within the concept of a circular economy, which is currently in its advent. It should be emphasized that such AD technological solutions are followed by challenges that limit the wider applicability, i.e., (i) the lack of markets for the digestate produced resulting from poor public perception; (ii) lack of recognition of AD and composting, especially regarding nutrient recirculation; (iii) purity of segregated biowaste streams; and (iv) requirements in capital investment.

Author Contributions: Conceptualization, methodology and paper writing, C.V.; Conceptualization, graphical illustration and validation of results C.A.; AD-related data, I.A.; Software runs and sensitivity analysis, A.V.M.; Case study related data, C.P.; Literature review, E.F.; Structure formulation, N.M. All authors have read and agreed to the published version of the manuscript.

Funding: This research was funded by the INTERREG IPA Cross Border Cooperation Programme CCI 2014 TC 16 I5CB 009 grant number SC032 (ZEFFIROS project).

Acknowledgments: The authors would like to wholeheartedly thank the experts that participated to the MCDA application in the study area.

Conflicts of Interest: The authors declare no conflict of interest.

References

1. European Investment Bank. Circular Economy Guide—Supporting the Circular Transition. January 2019. Available online: https://www.eib.org/en/publications/the-eib-in-the-circular-economy-guide (accessed on 10 March 2020).
2. Wautelet, T. *The Concept of Circular Economy: Its Origins and Its Evolution*; Positive ImpaKT: Windhof, Luxembourg, 2019.
3. Sariatli, F. Linear economy versus circular economy: A Comparative and analyzer study for optimization of economy for sustainability. *Visegr. J. Bioecon. Sustain. Dev.* **2017**, *1*, 31–34. [CrossRef]
4. Ellen MacArthur Foundation. *Towards the Circular Economy: Economic Business Rationale for an Accelerated Transition*; Ellen MacArthur Foundation: Cowes, UK, 2012.
5. Michailidou, A.V.; Vlachokostas, C.; Moussiopoulos, N. A methodology to assess the overall environmental pressure attributed to tourism areas: A combined approach for typical all-sized hotels in Chalkidiki, Greece. *Ecol. Indic.* **2015**, *50*, 108–119. [CrossRef]
6. European Commission. *Directive 2009/28/EC on the Promotion of the Use of Energy from Renewable Sources*; Official Journal of the European Union: Strasbourg, France, 2009.
7. Mihai, F.C.; Ingrao, C. Assessment of biowaste losses through unsound waste management practices in rural areas and the role of home composting. *J. Clean. Prod.* **2018**, *172*, 1631–1638. [CrossRef]
8. Bhatia, S.K.; Joo, H.S.; Yang, Y.H. Biowaste-to-bioenergy using biological methods—A mini-review. *Energy Convers. Manag.* **2018**, *177*, 640–660. [CrossRef]
9. Thomsen, M.; Seghetta, M.; Mikkelsen, M.H.; Gyldenkærne, S.; Becker, T.; Caro, D.; Frederiksen, P. Comparative life cycle assessment of biowaste to resource management systems—A Danish case study. *J. Clean. Prod.* **2017**, *142*, 4050–4058. [CrossRef]
10. Jensen, M.B.; Møller, J.; Scheutz, C. Comparison of the organic waste management systems in the Danish–German border region using life cycle assessment (LCA). *Waste Manag.* **2016**, *49*, 491–504. [CrossRef] [PubMed]
11. Huttunen, S.; Manninen, K.; Leskinen, P. Combining biogas LCA reviews with stakeholder interviews to analyse life cycle impacts at a practical level. *J. Clean. Prod.* **2014**, *80*, 5–16. [CrossRef]
12. Iakovou, E.; Vlachos, D.; Achillas, C.; Anastasiadis, F. Design of sustainable supply chains for the agrifood sector: A holistic research framework. *Agric. Eng. Int. CIGR J.* **2014**, *16*, 1–10.
13. Delgado, M.; López, A.; Cuartas, M.; Rico, C.; Lobo, A. A decision support tool for planning biowaste management systems. *J. Clean. Prod.* **2020**, *242*, 118460. [CrossRef]
14. Vea, E.B.; Romeo, D.; Thomsen, M. Biowaste valorisation in a future circular bioeconomy. *Procedia CIRP* **2018**, *69*, 591–596. [CrossRef]
15. Banias, G.; Achillas, C.; Vlachokostas, C.; Moussiopoulos, N.; Stefanou, M. Environmental impacts in the life cycle of olive oil: A literature review. *J. Sci. Food Agric.* **2017**, *97*, 1686–1697. [CrossRef] [PubMed]
16. Ross, T.; Soland, R. A multicriteria approach to the location of public facilities. *Eur. J. Oper. Res.* **1980**, *4*, 307–321. [CrossRef]
17. Vučijak, B.; Kurtagić, S.M.; Silajdžić, I. Multicriteria decision making in selecting best solid waste management scenario: A municipal case study from Bosnia and Herzegovina. *J. Clean. Prod.* **2016**, *130*, 166–174. [CrossRef]
18. Rezaei, J. A systematic review of multi-criteria decision-making applications in reverse logistics. *Transp. Res. Procedia* **2015**, *10*, 766–776. [CrossRef]
19. Soltani, A.; Hewage, K.; Reza, B.; Sadiq, R. Multiple stakeholders in multi-criteria decision-making in the context of Municipal Solid Waste Management: A review. *Waste Manag.* **2015**, *35*, 318–328. [CrossRef] [PubMed]

20. Achillas, C.; Moussiopoulos, N.; Karagiannidis, A.; Banias, G.; Perkoulidis, G. Use of multi-criteria decision analysis to tackle waste management problems: A literature review. *Waste Manag. Res.* **2013**, *31*, 115–129. [CrossRef]
21. Simões Gomes, C.; Nunes, K.; Xavier, L.H.; Cardoso, R.; Valle, R. Multicriteria decision making applied to waste recycling in Brazil. *Omega* **2008**, *36*, 395–404. [CrossRef]
22. Vego, G.; Kučar-Dragičević, S.; Koprivanac, N. Application of multi-criteria decision-making on strategic municipal solid waste management in Dalmatia, Croatia. *Waste Manag.* **2008**, *28*, 2192–2201. [CrossRef]
23. Queiruga, D.; Walther, G.; Gonzalez-Benito, J.; Spengler, T. Evaluation of sites for the location of WEEE recycling plants in Spain. *Waste Manag.* **2008**, *28*, 181–190. [CrossRef]
24. Michailidou, A.V.; Vlachokostas, C.; Moussiopoulos, N.; Maleka, D. Life Cycle Thinking used for assessing the environmental impacts of tourism activity for a Greek tourism destination. *J. Clean. Prod.* **2016**, *111*, 499–510. [CrossRef]
25. Achillas, C.; Vlachokostas, C.; Moussiopoulos, N.; Banias, G.; Kafetzopoulos, G.; Karagiannidis, A. Social acceptance for the development of a Waste-to-Energy plant in an urban area: Application for Thessaloniki, Greece. *Resour. Conserv. Recycl.* **2011**, *55*, 857–863. [CrossRef]
26. Morrissey, A.J.; Browne, J. Waste management models and their application to sustainable waste management. *Waste Manag.* **2004**, *24*, 297–308. [CrossRef] [PubMed]
27. Taelman, S.; Sanjuan-Delmás, D.; Tonini, D.; Dewulf, J. An operational framework for sustainability assessment including local to global impacts: Focus on waste management systems. *Resour. Conserv. Recycl.* **2019**, *2*, 100005. [CrossRef]
28. Soukopová, J.; Vaceková, G.; Klimovský, D. Local waste management in the Czech Republic: Limits and merits of public-private partnership and contracting out. *Util. Policy* **2017**, *48*, 201–209. [CrossRef]
29. European Commission. *Directive 2008/98/EC on Waste (Waste Framework Directive)*; Official Journal of the European Union: Strasbourg, France, 2008.
30. Achillas, C.; Vlachokostas, C.; Moussiopoulos, N.; Perkoulidis, G.; Banias, G.; Mastropavlos, M. Electronic waste management cost: A scenario based analysis for Greece (2011). *Waste Manag. Res.* **2011**, *29*, 963–972. [CrossRef] [PubMed]
31. Le Hesran, C.; Ladier, A.L.; Botta-Genoulaz, V.; Laforest, V. Operations scheduling for waste minimization: A review. *J. Clean. Prod.* **2019**, *206*, 211–226. [CrossRef]
32. Silva, S.; Alçada-Almeida, L.; Dias, L. Biogas plants site selection integrating Multicriteria Decision Aid methods and GIS techniques: A case study in a Portuguese region. *Biomass Bioenergy* **2014**, *71*, 58–68. [CrossRef]
33. Banias, G.; Achillas, C.; Vlachokostas, C.; Moussiopoulos, N.; Tarsenis, S. Assessing multiple criteria for the optimal location of a construction and demolition waste management facility. *Build. Environ.* **2010**, *45*, 2317–2326. [CrossRef]
34. Iakovou, E.; Moussiopoulos, N.; Xanthopoulos, A.; Achillas, C.; Michailidis, N.; Chatzipanagioti, M.; Koroneos, C.; Bouzakis, K.D.; Kikis, V. Multicriteria Matrix: A methodology for end-of-life management. *Resour. Conserv. Recycl.* **2009**, *53*, 329–339. [CrossRef]
35. Rousis, K.; Moustakas, K.; Malamis, S.; Papadopoulos, A.; Loizidou, M. Multi-criteria analysis for the determination of the best WEEE management scenario in Cyprus. *Waste Manag.* **2008**, *28*, 1941–1954. [CrossRef]
36. Hokkanen, J.; Salminen, P. Choosing a solid waste management system using multicriteria decision analysis. *Eur. J. Oper. Res.* **1997**, *98*, 19–36. [CrossRef]
37. Chatterjee, P.; Yazdani, M.; Chakraborty, S.; Panchal, D.; Bhattacharyya, S. (Eds.) Advanced Multi-Criteria Decision Making for Addressing Complex Sustainability Issues. In *Advances in Environmental Engineering and Green Technologies*, 1st ed.; IGI Global: Hershey, PA, USA, 2019.
38. Ishizaka, A.; Nemery, P. *Multi-Criteria Decision Analysis: Methods and Software*; Wiley: Hoboken, NJ, USA, 2013.
39. Al-Shemmeri, T.; Al-Kloub, B.; Pearman, A. Model choice in multicriteria decision aid. *Eur. J. Oper. Res.* **1997**, *97*, 550–560. [CrossRef]
40. Roy, B. Electre III: Algorithme de classement base sur une representation floue des preferences en presence de criteres multiples. *Cah. CERO* **1978**, *20*, 3–24.
41. Roy, B.; Bouyssou, D. *Aide Multicritere a la Decision: Methods et Cas*; Economica: Paris, France, 1993.

42. Spyridi, D.; Vlachokostas, C.; Michailidou, A.V.; Sioutas, C.; Moussiopoulos, N. Strategic planning for climate change mitigation and adaptation: The case of Greece. *Int. J. Clim. Chang. Strateg. Manag.* **2015**, *7*, 272–289. [CrossRef]
43. Vlachokostas, C.; Michailidou, A.V.; Matzaris, E.; Achillas, C.; Moussiopoulos, N. A multiple criteria decision-making approach to put forward tree species in urban environment. *Urban Clim.* **2014**, *10*, 105–118. [CrossRef]
44. Rogers, M.; Bruen, M. Choosing realistic values of indifference, preference and veto thresholds for use with environmental criteria within ELECTRE. *Eur. J. Oper. Res.* **1998**, *107*, 542–551. [CrossRef]
45. Roussat, N.; Dujet, C.; Mehu, J. Choosing a sustainable demolition waste management strategy using multicriteria decision analysis. *Waste Manag.* **2009**, *29*, 2–20. [CrossRef]
46. Vlachokostas, C.; Achillas, C.; Moussiopoulos, N.; Banias, G. Multicriteria methodological approach to manage urban air pollution. *Atmos. Environ.* **2011**, *45*, 4160–4169. [CrossRef]
47. Republic of Greece. Law 4513/2018 on Energy Communities. In *Official Government Gazette 9/A'/23.01.2018*; Hellenic Republic: Athens, Greece, 2018.
48. ZEFFIROS Project. Official Website. 2020. Available online: https://www.zeffirosproject.eu (accessed on 14 March 2020).
49. Michailidou, A.V.; Vlachokostas, C.; Moussiopoulos, N. Interactions between climate change and the tourism sector: Multiple-criteria decision analysis to assess mitigation and adaptation options in tourism areas. *Tour. Manag.* **2016**, *55*, 1–12. [CrossRef]
50. Haralambopoulos, D.; Polatidis, H. Renewable energy projects: Structuring a multi-criteria group decision-making framework. *Renew. Energy* **2003**, *28*, 961–973. [CrossRef]
51. Kourmpanis, B.; Papadopoulos, A.; Moustakas, K.; Kourmoussis, F.; Stylianou, M.; Loizidou, M. An integrated approach for the management of demolition waste in Cyprus. *Waste Manag. Res.* **2008**, *26*, 573–581. [CrossRef] [PubMed]
52. Kumar, A.; Samadder, S. Performance evaluation of anaerobic digestion technology for energy recovery from organic fraction of municipal solid waste: A review. *Energy* **2020**, *197*, 117253. [CrossRef]
53. Srisowmeya, G.; Chakravarthy, M.; Nandhini Devi, G. Critical considerations in two-stage anaerobic digestion of food waste—A review. *Renew. Sustain. Energy Rev.* **2020**, *119*, 109587. [CrossRef]
54. Wainaina, S.; Awasthi, M.K.; Sarsaiya, S.; Chen, H.; Singh, E.; Kumar, A.; Ravindran, B.; Awasthi, S.K.; Liu, T.; Duan, Y.; et al. Resource recovery and circular economy from organic solid waste using aerobic and anaerobic digestion technologies. *Bioresour. Technol.* **2020**, *30*, 122778. [CrossRef] [PubMed]
55. Pramanik, S.K.; Suja, F.B.; Zain, S.; Pramanik, B.K. The anaerobic digestion process of biogas production from food waste: Prospects and constraints. *Bioresour. Technol. Rep.* **2019**, *8*, 100310. [CrossRef]
56. Mc Cabe, B.; Schmidt, T. *Integrated Biogas Systems—Local Applications of Anaerobic Digestion towards Integrated Sustainable Solutions*; Murphy, J.D., Ed.; International Energy Agency: Cork, Ireland, 2018.
57. Al Seadi, T.; Stupak, I.; Smith, C.T. *Governance of Environmental Sustainability of Manure-Based Centralised Biogas Production in Denmark*; Murphy, J.D., Ed.; IEA Bioenergy: Cork, Ireland, 2018.
58. Holm-Nielsen, J.B.; Halberg, N.; Hutingford, S.; Al Seadi, T. Joint biogas plant. Agricultural advantages—Circulation of N, P and K. In *Report made for the Danish Energy Agency*, 2nd ed.; Danish Energy Agency: Copenhagen, Denmark, 1997.
59. Fagerström, A.; Al Seadi, T.; Rasi, S.; Briseid, T. *The Role of Anaerobic Digestion and Biogas in the Circular Economy*; Murphy, J.D., Ed.; IEA Bioenergy: Cork, Ireland, 2018.

energies

MDPI

Article

Forecasting of Day-Ahead Natural Gas Consumption Demand in Greece Using Adaptive Neuro-Fuzzy Inference System

Konstantinos Papageorgiou [1,*], Elpiniki I. Papageorgiou [2,3], Katarzyna Poczeta [4], Dionysis Bochtis [3,*] and George Stamoulis [5]

1 Department of Computer Science and Telecommunications, University of Thessaly, 35100 Lamia, Greece
2 Department of Energy Systems, Faculty of Technology, University of Thessaly-Geopolis, 41500 Larissa, Greece; elpinikipapageorgiou@uth.gr or e.papageorgiou@certh.gr
3 Center for Research and Technology—Hellas (CERTH), Institute for Bio-economy and Agri-technology (iBO), 57001 Thessaloniki, Greece
4 Department of Information Systems, Kielce University of Technology, 25-314 Kielce, Poland; k.piotrowska@tu.kielce.pl
5 Department of Electrical and Computer Engineering, University of Thessaly, 38221 Volos, Greece; georges@e-ce.uth.gr
* Correspondence: konpapageorgiou@uth.gr (K.P.); d.bochtis@certh.gr (D.B.)

Received: 31 March 2020; Accepted: 4 May 2020; Published: 7 May 2020

Abstract: (1) Background: Forecasting of energy consumption demand is a crucial task linked directly with the economy of every country all over the world. Accurate natural gas consumption forecasting allows policy makers to formulate natural gas supply planning and apply the right strategic policies in this direction. In order to develop a real accurate natural gas (NG) prediction model for Greece, we examine the application of neuro-fuzzy models, which have recently shown significant contribution in the energy domain. (2) Methods: The adaptive neuro-fuzzy inference system (ANFIS) is a flexible and easy to use modeling method in the area of soft computing, integrating both neural networks and fuzzy logic principles. The present study aims to develop a proper ANFIS architecture for time series modeling and prediction of day-ahead natural gas demand. (3) Results: An efficient and fast ANFIS architecture is built based on neuro-fuzzy exploration performance for energy demand prediction using historical data of natural gas consumption, achieving a high prediction accuracy. The best performing ANFIS method is also compared with other well-known artificial neural networks (ANNs), soft computing methods such as fuzzy cognitive map (FCM) and their hybrid combination architectures for natural gas prediction, reported in the literature, to further assess its prediction performance. The conducted analysis reveals that the mean absolute percentage error (MAPE) of the proposed ANFIS architecture results is less than 20% in almost all the examined Greek cities, outperforming ANNs, FCMs and their hybrid combination; and (4) Conclusions: The produced results reveal an improved prediction efficacy of the proposed ANFIS-based approach for the examined natural gas case study in Greece, thus providing a fast and efficient tool for utterly accurate predictions of future short-term natural gas demand.

Keywords: neuro-fuzzy; ANFIS; neural networks; soft computing; fuzzy cognitive maps; energy forecasting; natural gas; prediction

1. Introduction

The increasing technological advancements and the rapid global population growth have led to a remarkable increase in energy consumption all over the world and especially in the developed and developing countries. After all, energy consumption is an index of a society's economical welfare and represents the economic development of a city or country [1]. Due to this unexpected boost in energy consumption over the past few decades, the need for energy demand management became crucial for achieving economic success that will result in self-sufficiency and economic development [2]. Thus, energy consumption forecasting is essential, as it predicates an energy efficient policy, the optimization of usage [3] and energy supplies optimum management. Even though a variety of methods have been investigated for energy demand forecasting, this is not an easy task, as it is affected by uncertain exogenous factors such as weather, technological development and government policies [4].

Among all energy resources, natural gas (NG) has received the largest increase in consumption lately [5], mostly due to its popularity as a clean energy source, with respect to environmental concerns. In particular, this energy source is characterized by low-level emissions of greenhouse gases in comparison with other non-renewable energy sources [6,7] and is considered as the cleanest-burning fossil fuel [8–11]. One fifth of the world's primary energy demand is covered by NG [7] and is linked to industrial production, transportation, health, agricultural output and household use.

According to the sharp increase in NG consumption, the forecasting of NG consumption has attracted great attention since it is essential for project planning, gas imports, tariff design, optimal scheduling of the NG supply system [12], indigenous production, infrastructures planning and cost reduction at different levels [7]. Moreover, some noteworthy factors like the need for distribution planning, especially in residential areas, the increasing demand of NG and the restricted NG network in many countries, make consumption forecasting on an hourly, daily, weekly, monthly or yearly basis highly important [13].

Especially in national energy strategy, NG demand forecasting is of high importance and can help policymakers all over the world to choose certain strategies in this direction. The development of systems that model NG consumption could assist in good government policymaking. Over the past decade, there has also been a significant increase in NG consumption in Greece and so the prediction of demand has become crucial accordingly. In particular, there is a need for further distribution planning in the Greek territory, especially in residential areas and in cases of high demand, when the accumulation ability of the network itself is decreased [13].

Selecting appropriate load forecasting methods is the most important step. Various models have been proposed in the field of energy forecasting so far, and can be divided into three groups: (i) traditional statistical models such as regression analysis, time series methods and ARIMA [14]; (ii) artificial intelligence (AI)-based methods such as wavelet analysis, artificial neural network (ANN) methods, neuro-fuzzy, machine learning, gray theory prediction; and (iii) hybrid models, which are a combination of various forecasting methods [15–23]. As regards the forecasting period horizon, prediction of demand is classified into three categories: short-term, medium-term and long-term forecasting. Short-term forecasting is used for hourly, daily and weekly demand predictions and in regard to NG, it is oriented mainly in system management and balancing, storage capacities optimization, as well as in making optimum purchasing and operating decisions [24,25]. Medium-term forecasting deals with seasonal demand prediction (one to several months) and mainly plans the fuel purchases, while long-term forecasts (more than a year ahead) aim to develop the power supply and delivery system [26,27].

ANN, genetic algorithm (GA) and fuzzy inference systems (FIS), as artificial intelligence (AI) techniques, are among those methods that are often used in the energy domain and specifically for energy demand forecasting, due to their high flexibility and reasonable estimation and prediction ability. From the relevant literature, there have been several attempts in energy forecasting by ANNs, like those in [28–32]. ANNs have been applied to forecast electric energy consumption in Saudi Arabia [33], energy consumption of a passive solar building [34], energy consumption of the Canadian residential sector [35], the peak load of Taiwan [36], while ANNs have been further explored in [37–41] for short-term load forecasting. An abductive network machine learning for predicting monthly electric energy consumption in domestic sector of Eastern Saudi Arabia was proposed in [42]. Moreover, support vector machines (SVMs) and genetic algorithms (GA) were explored in [29,30,43] to predict electricity load. In [44], SVMs coupled with empirical mode decomposition were used to perform long term load forecasting. GAs have been used in [45] and [46] to estimate Turkey's energy demand and electricity demand in the industrial sector, respectively. In addition, Azadeh et al. have proposed the integration of GAs and ANNs to estimate and predict electrical energy consumption [47]. A number of literature reviews have also been performed regarding various energy fields. For example, in [48], a review of the conventional methods and AI methods for electricity consumption forecasting was provided, while in [49] the strengths, shortcomings, and purpose of numerous AI-based approaches in the energy consumption forecasting of urban and rural-level buildings were discussed. In [50], a review about conventional models, including time series models, regression models and gray models was conducted with respect to energy consumption forecasting. Moreover, an overview of AI methods in short term electric load forecasting area was discussed in [51].

1.1. Indicative Related Work on AI Applied in Natural Gas Consumption Forecasting

On the other hand, there are many studies where different AI methods like neural networks, neuro-fuzzy and other ANN topologies have been investigated and applied in NG demand forecasting [52–60]. In this context, ANNs were extensively used in [21,54,57,60–69] to investigate short-term NG forecasts, while in [56] different types of ANN algorithm were explored to forecast gas consumption for residential and commercial consumers in Istanbul, Turkey. ANNs were also used in [70] for the daily and weekly prediction of NG consumption of Siberia, using historical temperature and NG consumption data, in [71] for NG output prediction of USA until 2020, as well as in [72] for the prediction of NG consumption and production in China from 2008 to 2015, applying the grey theory along with NNs. Furthermore, a combination of ANNs was applied in [54,55] for the prediction of NG consumption at a citywide distribution level.

More recent studies presented numerous techniques for NG demand forecasting, including computational intelligence-based models (ANNs), fuzzy logic and support vector machines [12,73–76]. In this sense, a combination of recurrent neural network and linear regression model was used in [77] to generate forecasts for future gas demand, whereas a multilayered perceptron (MLP) neural network was deployed in [78] to estimate the next day gas consumption. A day-ahead forecast was also examined in [79] by developing a functional autoregressive model with exogenous variables (FARX). Moreover, machine learning tools such as multiple linear regression (MLR), ANN and support vector regression (SVR) were devised in [80] to project NG consumption in the province of Istanbul, as well as in [77] to forecast the residential NG demand in the city of Ljubljana, Slovenia. When considering AI methods, self-adapting intelligent grey models were also deployed for forecasting NG demand, as in [81,82].

Regarding neural network algorithms, the multilayer perceptron and the radial basis function network with different activation functions were trained and tested in [21,63,65], while the authors of [66] used a multilayer perceptron algorithm for neural network and compared this model with two time series models. In addition, Taspinar et al. explored daily gas consumption forecasting through different methods including the seasonal autoregressive integrated moving average model with exogenous inputs (SARIMAX), multi-layer perceptron ANN (ANN-MLP), ANN with radial basis functions (ANN-RBF), and multivariate ordinary least squares (OLS) [63]. Different sets of AI methods were also implemented in the following studies regarding NG consumption. ANNs with linear regression models were used in [83] for daily prediction, while ANN and Fuzzy ANN models were investigated in [84] regarding consumption in a certain region of Poland. Finally, it is worth mentioning that there has been similar research in [12], which proposed the hybrid wavelet-ANFIS/NN model to compute day-ahead forecasts for 40 distribution nodes in the national NG system of Greece.

1.2. Related Work on ANFIS in Energy Consumption Forecasting

There are plenty of works in the relevant literature regarding the application of the ANFIS model in forecasting energy consumption. As regards the electricity domain, ANFIS model was applied to forecast annual regional load in Taiwan [85] and annual demand in Turkey [86], showing in both cases that the results are good and the ANFIS model performed better than regression, neural network and fuzzy hybrid systems. ANFIS was also used in [87] for short-term electricity demand forecasting, using weekly electricity load data, as well as in [88], to estimate possible improvement of electricity consumption. Also, for electricity load forecasting, ANFIS was used in [89] to highlight its superiority to the ANN model, while it was furthermore applied in the field of transportation, forecasting the corresponding energy demand for the years 2010 to 2030, in the country of Jordan, revealing the efficiency of the examined model. Another study regarding the energy domain, where the ANFIS model was applied, is that of [90]. A long-term prediction of oil consumption was studied, further examining the interrelationship between oil consumption and economic growth in Turkey, for the years 2012 to 2030.

1.3. Related Work on ANFIS in Natural Gas Consumption Forecasting

Casting a view on the literature that refers to NG consumption forecast, the authors came across only one study that devised solely an ANFIS model. Specifically, ANFIS was used in [91] in order to estimate the daily NG demand in Iran, which actually used an extremely small dataset of historical data for both testing and training (December 2007–June 2008). Models trained on a small dataset tend to overfit, which results in high variance and very high error on a test set, producing inaccurate results. In this case, the predicting error decreases monotonically with the size of training set [92]. The rest of the studies dealt with approaches that combine ANFIS with other methods. For example, in [21], statistical time series analysis along with ANN and ANFIS methods were applied in order to predict weekly NG consumption in Turkey. Moreover, an ANFIS-fuzzy data envelopment analysis (FDEA) was developed in [93] for long-term NG consumption forecasting and analysis. In this study, 104 ANFIS were constructed and tested and six models were proposed to forecast annual NG consumption. The same approach was proposed in [94] for accurate gas consumption estimation in South America with noisy inputs. An ANFIS-stochastic frontier analysis (ANFIS-SFA) approach was formulated in [95] for long-term NG consumption prediction and analysis. Three patterns of the hybrid ARIMA–ANFIS model were tested in [2] to predict the annual energy consumption in Iran, using a set of data like population, GDP, export and import. Finally, a hybrid model of adaptive neuro fuzzy inference system and computer simulation for the prediction of NG consumption was developed in [96].

1.4. Related Work on Fuzzy Cognitive Maps (FCMs) in Energy and Natural Gas Consumption Forecasting

Moreover, other soft computing techniques, like evolutionary fuzzy cognitive maps (FCMs) have been applied for the modeling and prediction of time series problems. The dynamic modeling structure of FCMs inheriting the learning capabilities of recurrent neural networks works properly for modeling and time series prediction. Salmeron and Froelich further investigated the applicability of FCMs in univariate time series prediction by proposing an FCM simplification approach with the removal of nodes and weights [97]. Regarding the task of multivariate time series prediction, Froelich and Salmeron proposed a nonlinear predictive model based on an evolutionary algorithm for learning fuzzy grey cognitive maps [98], while Papageorgiou et al. [99] and Poczeta et al. [100] applied a new type of evolutionary FCM enhanced with the Structure Optimization Genetic Algorithm (SOGA) in energy for electricity load forecasting. Through the SOGA algorithm, an FCM model can automatically be constructed by taking into consideration any available historical data. A two-stage prediction model for multivariate time series prediction, based on the efficient capabilities of evolutionary FCMs and enhanced by structure optimization algorithms and ANNs, was introduced in [101]. In the first stage of the prediction model, SOGA-FCM was applied for selecting the most significant concepts and defining the relationships between them. Next, that model was fed into the second stage to define the initial features and weights of the training ANN. This generic prediction approach was applied in four common prediction problems, one of which dealt with electric power consumption.

In [102], Poczeta and Papageorgiou conducted a preliminary study on implementing FCMs with ANNs for NG prediction, showing for the first time the capabilities of evolutionary FCMs in this domain. Furthermore, the research team in [13] recently contacted a study for time series analysis devoted to NG demand prediction in three Greek cities, implementing an efficient ensemble forecasting approach through combining ANN, real coded genetic algorithm (RCGA)-FCM, SOGA-FCM, and hybrid FCM-ANN. In this research study, the advantageous features of intelligent methods through an ensemble to multivariate time series prediction in NG demand forecasting are explored.

1.5. Research Gap and the Novelty of This Study

Based on the reported literature survey and reviews [7,15], regarding the application of ANN-based and hybrid forecasting methods, a research gap has been identified in the field of NG day-ahead demand prediction. The observed gap mainly refers to the lack of model simplicity and flexibility, the insufficient exploration of certain modelling aspects, and inadequacy to cope with the inherent fuzziness in data handling. Most of these forecasting methods need a large dataset to be trained and a relatively large number of features to make accurate predictions. Furthermore, they are complex in their structure, time consuming and difficult to be used by non-experienced AI users. There has been hardly any research on successfully applying the ANFIS technique on the field of NG demand prediction, having performed a deep exploration process for determining the best model configuration, thus producing a highly accurate model with generalization capabilities.

Considering the aforementioned limitations, this work aims to fill the observed research gap and seeks to develop an easy to use, robust and flexible ANFIS model, which is at the same time fast, simple in structure and able to cope with fuzziness. More specifically, the proposed ANFIS architecture uses as model's inputs the most important and commonly used input variables according to the literature [7,15], such as day, month and daily average temperature, along with past NG consumption data. Moreover, a relatively large dataset was used for both testing and training the model, resulting in a systematic improvement of the model's predictive accuracy [92]. The current work pays great attention to the generalization of the proposed method and tries to properly evaluate the model's generalization capabilities in two ways: (i) by applying the model on a city level, where 10 different cities were properly examined, and (ii) by carrying out an exploration process, where 94 different models' configuration sets were examined for each one of the cities that participated in this research.

To sum up, the innovations offered in this paper are as follows, highlighting the contribution of this work to the research community:

- The creation and demonstration of a simple, fast, robust ANFIS prediction tool to forecast NG demand using historical time series data. The proposed model is characterized by high flexibility, especially in large datasets, easiness of use and low execution time requirements.
- The rigorous ANFIS fine-tuning for determining the most appropriate architecture for an enhanced prediction performance.

1.6. Aim of This Research Work

The motivation of this work is to propose an ANFIS-based forecasting approach with generalization capabilities for short-term (day-ahead) city-level NG prediction in Greek areas. Also, a comparative analysis is conducted, applying ANNs, evolutionary FCMs and hybrid combinations of them on the same dataset to show the capabilities of the proposed ANFIS architecture.

The objectives of the present paper regarding NG demand forecasting, are briefly summarized in the following:

(a) To develop a robust ANFIS model to provide accurate short-term forecasts for a number of cities in Greece, using a relatively large dataset. At the same time, the authors perform model fine-tuning that can lead to high accuracy in most distribution points. The proposed model is characterized by high flexibility, easiness of use and low execution time requirements.

(b) To apply FCMs, ANNs and hybrid combinations of them to forecast NG demand in the same dataset, since these approaches have been proved as efficient techniques for NG demand forecasting according to the relevant literature.
(c) To assess the performance of these soft computing methods in terms of prediction accuracy using well-known evaluation metrics.
(d) To compare forecasting accuracy results of the proposed approach with those of the other soft computing and ANN methods that were examined, and finally decide on which model offers the best forecasting accuracy.

The outline of this paper is as follows. Section 2 describes the datasets of NG demand, collected for 10 Greek distribution points, as well as the proposed methodology of ANFIS for NG demand prediction using a well-defined set of evaluation performance metrics. Section 3 presents the results of the investigated ANFIS architectures. In the same section, a comparative analysis with other traditional neural networks and soft computing methods was performed for the same dataset. The discussion of results, which is also included in Section 3, presents the main outcomes of a meticulous ANFIS exploration analysis, along with ANFIS advantageous features. These are compared with ANNs and FCMs, and their overall contribution in NG forecasting is presented. Section 4 summarizes the paper, presenting future challenges in energy demand forecasting and highlighting further research directions.

2. Materials and Methods

This study aims to develop an ANFIS architecture capable of forecasting short-term NG consumption demand of the 10 main cities in Greece using the dataset that was provided by the Hellenic Gas Transmission System Operator S.A. (DESFA) [103]. The developed ANFIS approach deployed the aforementioned dataset along with other variables like the average daily temperature data for all the examined cities to accomplish forecasting. The results produced were further compared with those calculated by ANN and other soft computing techniques like FCM and hybrid-ANN to prove the prediction performance of the ANFIS prediction tool. Details on the dataset and its features, as well as the proposed methodology, are provided below. MATLAB M-file environment version 9.3.0.71 (R2017b) was used to program ANFIS networks and develop ANFIS models.

2.1. Dataset

The dataset covers ten different prediction datasets of historical data referring to ten cities all over Greece (Alexandroupoli, Athens, Drama, Karditsa, Larissa, Markopoulo, Serres, Thessaloniki, Trikala and Volos) and was linked to the values of gas demand for eight (8) previous years, in total. It should be mentioned that the time period for each dataset (city) was not the same in duration and did not correspond to the same years of data with all the other datasets collected. Table 1 depicts the duration in years that is linked to each dataset collected and used in this case study. The historical datasets for 15 Greek cities were initially provided by the NG grid company of Greece, DESFA, which is responsible for the operation, management, exploitation and development of the Greek NG system and its interconnections. However, the authors, after thoroughly reviewing the available datasets, decided to include only 10 out of 15 cities in their case study, since these datasets contained less outliers and missing values than the other 5 datasets that were finally rejected, for data consistency purposes. For the datasets that were finally included in this work, a preliminary preprocessing phase was performed, where the insignificant outliers were removed, and any missing values were substituted with the average real value of the previous two days demand. The real data that were used for ANFIS modeling, performance evaluation and comparison with other popular forecasting methods were then split into training and testing samples. For all cities, the last year of each dataset (from November 2017 up to October 2018) was devoted to testing, whereas the rest of the years were used for training the developed ANFIS model.

Table 1. Time period referred to in each time-series dataset for all cities.

City	Time Period of the Examined Data	City	Time Period of the Examined Data
Alexandroupoli	2/2013–10/2018	Markopoulo	3/2010–10/2018
Athens	3/2010–10/2018	Serres	6/2013–10/2018
Drama	9/2011–10/2018	Thessaloniki	3/2012–10/2018
Karditsa	5/2014–10/2018	Trikala	9/2012–10/2018
Larissa	3/2010–10/2018	Volos	3/2010–10/2018

In order to properly forecast day-ahead NG consumption demand of Greece, the proper number and type of input parameters should be selected. So, five factors were carefully considered as input parameters and the amount of one-day-ahead NG consumption demand of each distribution point was the output parameter. The prediction model was based on observations of past NG consumption, weather data, and calendar indicators, which are all among the most important input variables for prediction of NG consumption [15]. In particular, the dataset contains historical data of NG consumption of each city's distribution point, the daily average temperature of the area in Celsius degrees, a month indicator and a day indicator. As regards the previous NG consumption data, these are linked to two different input variables: demand of a day before and current day demand. The temperature data are obtained by the nearest to the distribution gas point meteorological station. Concerning the calendar indicators (month and day), they need to undergo certain data form preprocessing before their use. Specifically, two different input indicators need to be considered for each one of the two variables. We define k = 1,2, ... , 12 as the month index (1 January, 2 February, ... , 12 December) and l = 1,2, ... , 7 as the day index (1 Monday, 2 Tuesday, ... 7 Sunday). Following the coding procedure as presented in [104], the index for the month is scaled to the range [1/12, 1] in which the months of the year from January to December take successive values of the scaled index. That is, January has the value of 1/12 and December the value of 1. Similarly, the days of the week take successive values in the scaled range [1/7, 1], in which Monday and Sunday take the values of 1/7 and 1, respectively. All these parameters constituting the actual recorded data are briefly presented in Table 2.

Table 2. Input and output parameters.

Type	Parameter	Unit
Input	Demand of a day before	MWh
Input	Current day demand	MWh
Input	Daily average temperature	Celsius degrees
Input	Month indicator	K = 1/12, 2/12, ... , 1
Input	Day indicator	l = 1/7, 2/7, ... , 1
Output	A day ahead NG demand	MWh

All data that compose the investigated dataset underwent a normalization process. This was necessary because all entries needed to have the same limited range of values so the model produces meaningful results [105].

The algorithm that was used for data normalization is the Min-Max, which scales the values of the dataset linearly over a specific range. As described in previous works [13,105], each variable was normalized to [0,1] before the forecasting model was applied. The normalized variable took its original value when the testing phase was implemented. Data normalization was carried out mathematically, as follows:

$$x_i^{(new)} = \frac{x_i - x^{(min)}}{x^{(max)} - x^{(min)}}, \ \forall i = 1, 2, \ldots, N \tag{1}$$

where $x^{(new)}$ is the normalized value of the variable x, and $x^{(min)}$ and $x^{(max)}$ are, respectively, the minimum and maximum values of the concerned variable x.

2.2. Methods

2.2.1. Adaptive Neuro-Fuzzy Inference System (ANFIS)

The adaptive neuro-fuzzy inference system (ANFIS) uses an architecture that is based on both ANN and fuzzy logic principles and takes advantage of the benefits of both in a single framework. It can be described by the fuzzy "IF-THEN" rules from the Takagi and Sugeno (TS) type [106] as follows:

$$\begin{gathered} R_i : if\ x_1 = A_{i,1}\ and \ldots and\ x_k = A_{i,k} \\ then\ y_i = b_{i,0} + b_{i,1}x_1 + b_{i,2}x_2 + \ldots + b_{i,k}x_k \end{gathered} \tag{2}$$

where $A_{i,k}$ is the membership function associated with input variables x_k and n is the number of inputs.

A typical ANFIS network is a five-layer structure consisting of the fuzzy layer, the product layer, the normalized layer, the de-fuzzy layer and the total output layer [3,107,108], as depicted in Figure 1.

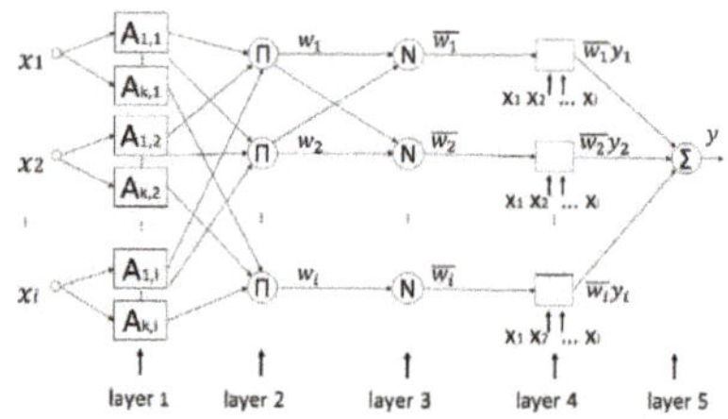

Figure 1. The TSK ANFIS architecture [107].

In the first layer, every node i represents a linguistic label and is described by the following membership function, as given in Equation (3).

$$A_{i,k}(x_r) = e^{-\left(\frac{(x_r - v_{i,k})}{\sigma_{i,k}}\right)^2},\ for\ r = 1,\ 2, \ldots, i \tag{3}$$

where $A_{i,k}$ is the membership function which is considered to be Gaussian and is described by the center v and the spread σ.

In the second layer, the firing strength of the rule is computed using multiplicative operator, as presented in Equation (4). Firing strength is the weight degree of the IF-THEN rule and determines the shape of the output function for that rule.

$$w_i = \prod_{k=1}^{n} A_{i,k}(x_k) \tag{4}$$

In the third layer, the i-th node calculates the ratio of the i-th rule's firing strength to the sum of the firing strength of all rules. This is the normalization layer which normalizes the strength of all rules and the output of each node is given by Equation (5).

$$\overline{w}_i = \frac{w_i}{\sum_{i=1} w_i} \tag{5}$$

In the fourth layer, each node is an adaptive node with a function given by Equation (6). In this layer, each node calculates a linear function where its coefficients are adapted by using the error function of the multilayer feed-forward neural network.

$$\overline{y}_i = \overline{w}_i y_i \tag{6}$$

In the fifth layer, there is only a fixed node indicated as the sum of the net outputs of the nodes in Layer 4. It computes the overall output as the sum of all incoming inputs and is expressed by Equation (7).

$$y = \sum_i \overline{y}_i \quad (7)$$

ANFIS uses a hybrid learning algorithm to train the model. The back-propagation algorithm is used to train the parameters in Layer 1, whereas a variation of least-squares approximation or back-propagation algorithm is used for training the parameters of the fourth layer [108,109].

2.2.2. Proposed ANFIS Architecture Applied in Natural Gas Consumption Forecasting

In order to develop an efficient ANFIS model for NG demand forecasting, the authors needed to follow a certain process regarding the design of model's architecture as well as an exploration process that will properly configure the input and training parameters of the examined model. Priority was given to the definition of the FIS architecture before the training of the network [110]. Among various fuzzy inference system (FIS) models, the Sugeno fuzzy model is the most widely used because of its higher interpretability and computational ability, that includes embedded optimal and adaptive techniques [111]. In order to create a fuzzy rule, the input space needs first to be divided. Two methods are used to divide space, comprised by input variables: the grid partitioning method and the subtractive clustering method. The main difference between these two functions refers to the way the partition of the input space is created.

In grid partitioning [109], the input space is divided into a grid like structure without overlapping parts. Grid partitioning performs partitioning of the input space using all possible combinations of membership functions of each variable. This method is used when the number of input variables is small. For example, for 10 input variables and two membership functions for each input variable, then the input space is divided into $2^{10} = 1024$ specific areas, representing one rule for each specific area, and the total number of rules is 1024, which is a very complicated structure. Therefore, the grid partitioning method is mainly used when the number of input variables is small.

On the other hand, the subtractive clustering method divides the input space into appropriate clusters, even if the user does not specify their number. If the size of the cluster becomes small, then the number of clusters increases, thus increasing the number of fuzzy rules. A rule is created for each cluster, whereas different values for parameters, like range of influence, squash factor, accept ratio and reject ratio, need to be explored for determining an efficient architecture, which will keep the balance between the total number of ANFIS parameters and the total number of rules.

Considering the above specifications, the authors used the Grid partition option to define the FIS architecture due to its simplicity, less time-consuming performance as well as it can easily explore the number and type of membership function (MF). In this stage, the number and type of membership functions of each input variable, along with the rules and values of parameters that belong to these functions, were determined using the option of Grid partition.

When implementing an ANFIS architecture, researchers should have in mind that there is one main restriction: the number of input variables. When these are more than five, then the number of the IF-THEN rules and the computational time also increase, hindering ANFIS to model output with respect to inputs [110]. Thus, in this study, five variables were chosen as input parameters, i.e., month, day, temperature, demand of a day before and demand of current day. As described above, a day-ahead consumption demand was selected as the output variable whose value can be produced by choosing between the option of linear or constant type of MF.

Finding the most efficient ANFIS architecture is a demanding task and entails a rigorous exploration process. Since our concern focuses on the increment of network's accuracy and decrement of the errors, five necessary configurations should be considered in this direction: (i) the number of membership functions, (ii) types of MF (triangular, trapezoidal, bell-shaped, Gaussian and sigmoid), (iii) types of output MF (constant or linear), (iv) optimization methods (hybrid or back propagation) and (v) the number of epochs [112]. For the convenience of readers, these steps are visually represented in the flowchart in Figure 2.

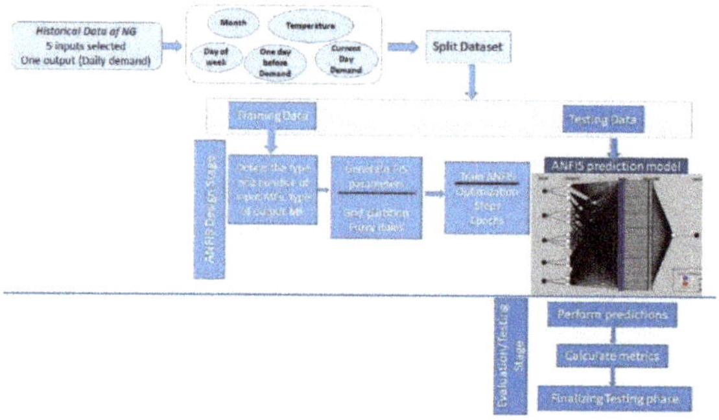

Figure 2. Flowchart of the proposed adaptive neuro-fuzzy inference system (ANFIS) methodology.

The aforementioned set of configurations needs to be deployed in order to generate FIS and next to train the ANFIS model. Accordingly, the dataset that included the five input variables (i.e., month, day, temperature, demand of a day before, demand of current day) was selected to determine the only output (day-ahead demand). Initially, the training dataset was loaded in the ANFIS tool, as shown in Figure 3a. The next step was the design of the neuro-fuzzy model using the option "Generate FIS". The Grid partition option was selected according to the description above (see Figure 3a). These two settings, concerning the fuzzy input variables along with their membership functions, are the most important parts to design the ANFIS. An example of selecting the number and type of MFs is illustrated in Figure 3b.

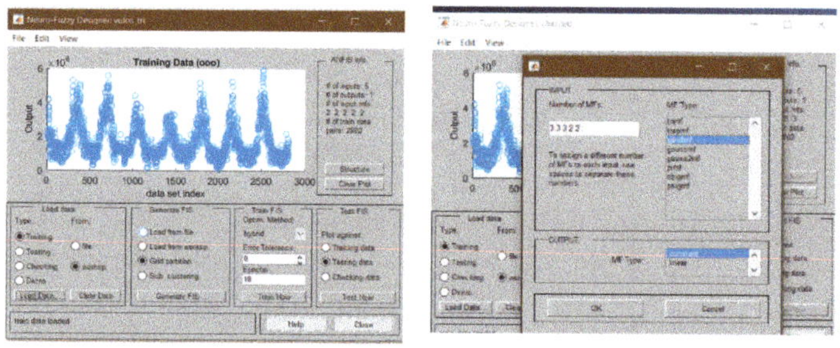

(**a**) (**b**)

Figure 3. Screenshots regarding (**a**) the training dataset configuration, (**b**) the number and type of MF configuration.

The number and type of membership function were assigned to the input parameters following the trial-and-error approach. The different types of MF that are offered by the MATLAB ANFIS editor include the triangular, trapezoidal, generalized bell (Gbell), Gaussian curve, Gaussian combination, difference between two sigmoid functions and product of two sigmoid functions (see Figure 3b). Regarding the type of output MFs, in the Sugeno-type fuzzy system, there are two options: a constant-type conclusion or a linear-type conclusion function. In the case of linear function, the output y is defined as:

$$y = k_0 + k_1 * x_1 + k_2 * x_2 + \ldots + k_n * x_n \qquad (8)$$

where $x_1, x_2, \ldots, x_n$ are the n inputs. In this case, ANFIS needs to define k_0, k_1, k_2 up to k_n, and it is very time consuming to efficiently calculate the outputs when a large number of parameters are considered. On the other hand, when a constant MF is selected, the algorithm needs to define only one parameter to provide a reliable forecasted value. Thus, the computational time is really low.

The selected configuration also includes the hybrid optimization method, while the number of epochs selected to train the model was between 10 and 50. The hybrid optimization method uses the back propagation learning algorithm for parameters associated with input MF and the least-square estimation algorithm for parameters associated with output MF; thus, it was selected as the most proper one [113]. Various sets of ANFIS configurations are presented in Table 3, regarding different sets of number for MFs, as considered by the authors of this work.

Table 3. Different configurations of the selected ANFIS architectures regarding constant output membership function (MF).

ANFIS Run	Type of Input MF	Number of MFs	Type of Output MF	Number of Epochs	Learning Method
1	trimf	2-2-2-2-2	Constant	10	Hybrid
2	trapmf	2-2-2-2-2	Constant	10	Hybrid
3	gbellmf	2-2-2-2-2	Constant	10	Hybrid
4	Gaussmf	2-2-2-2-2	Constant	10	Hybrid
5	Gauss2mf	2-2-2-2-2	Constant	10	Hybrid
6	pimf	2-2-2-2-2	Constant	10	Hybrid
7	dsigmf	2-2-2-2-2	Constant	10	Hybrid
8	psigmf	2-2-2-2-2	Constant	10	Hybrid
9	trimf	2-2-3-3-3	Constant	10	Hybrid
10	trapmf	2-2-3-3-3	Constant	10	Hybrid
11	gbellmf	2-2-3-3-3	Constant	10	Hybrid
12	Gaussmf	2-2-3-3-3	Constant	10	Hybrid
13	Gauss2mf	2-2-3-3-3	Constant	10	Hybrid
14	pimf	2-2-3-3-3	Constant	10	Hybrid
15	dsigmf	2-2-3-3-3	Constant	10	Hybrid
16	psigmf	2-2-3-3-3	Constant	10	Hybrid
17	trimf	3-3-3-2-2	Constant	10	Hybrid
18	trapmf	3-3-3-2-2	Constant	10	Hybrid
19	gbellmf	3-3-3-2-2	Constant	10	Hybrid
20	Gaussmf	3-3-3-2-2	Constant	10	Hybrid
21	trimf	3-3-3-3-3	Constant	10	hybrid
22	trimf	3-3-3-3-3	Constant	10	backpropa
23	trapmf	3-3-3-3-3	Constant	10	hybrid
24	trapmf	3-3-3-3-3	Constant	10	backpropa
25	gbellmf	3-3-3-3-3	Constant	10	hybrid
26	gbellmf	3-3-3-3-3	Constant	10	backpropa
27	trimf	3-3-3-3-3	Constant	30	hybrid
28	trimf	3-3-3-3-3	Constant	50	hybrid
29	trapmf	3-3-3-3-3	Constant	30	hybrid
30	trapmf	3-3-3-3-3	Constant	50	hybrid
31	gbellmf	3-3-3-3-3	Constant	30	hybrid
32	gbellmf	3-3-3-3-3	Constant	50	hybrid
33	trimf	3-3-4-4-4	Constant	10	hybrid
34	trimf	3-3-5-5-5	Constant	10	hybrid
35	trapmf	3-3-4-4-4	Constant	10	hybrid
36	trapmf	3-3-5-5-5	Constant	10	hybrid
37	gbellmf	3-3-4-4-4	Constant	10	hybrid
38	gbellmf	3-3-5-5-5	Constant	10	hybrid
39	gaussmf	3-3-3-3-3	Constant	10	hybrid
40	gaussmf	3-3-4-4-4	Constant	10	hybrid
41	gaussmf	3-3-5-5-5	Constant	10	hybrid
42	gauss2mf	3-3-3-3-3	Constant	10	hybrid
43	gauss2mf	3-3-4-4-4	Constant	10	hybrid
44	gauss2mf	3-3-5-5-5	Constant	10	hybrid
45	pimf	3-3-3-3-3	Constant	10	hybrid
46	pimf	3-3-4-4-4	Constant	10	hybrid
47	pimf	3-3-5-5-5	Constant	10	hybrid

Regarding the output MFs, constant and linear MF were accordingly investigated after certain numbers of experiments conducted. From these experiments and for the linear output, it was observed that the number of rules increases significantly, as well as the computational time, even in the case of problems with a small number of inputs (see Table A1 in Appendix A). Thus, the linear type was not considered as an appropriate parameter of output MF since it is extremely time consuming. In this context, the trial-and-error approach was followed for the selection of the input-output type of MFs. Figure 4 illustrates an indicative ANFIS model, which was constructed with the following configuration set: 3-3-3-2-2, gbell MF, constant output MF, 10 epochs, hybrid.

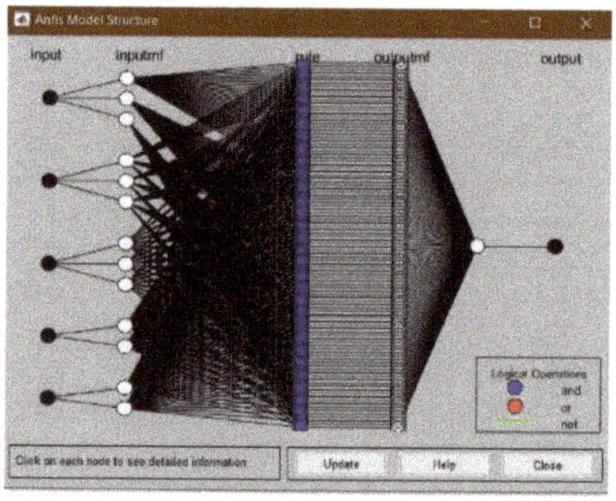

Figure 4. Screenshot of the constructed ANFIS model.

2.2.3. Testing and Evaluation

The testing process for the ANFIS model was accomplished by using the testing data, which were completely unknown to the model. The predictor makes predictions on each day and finally compares the calculated predicted value with the real value. For example, considering the city of Volos, the predicted values that are illustrated in red in Figure 5 are compared with the real values (in blue color).

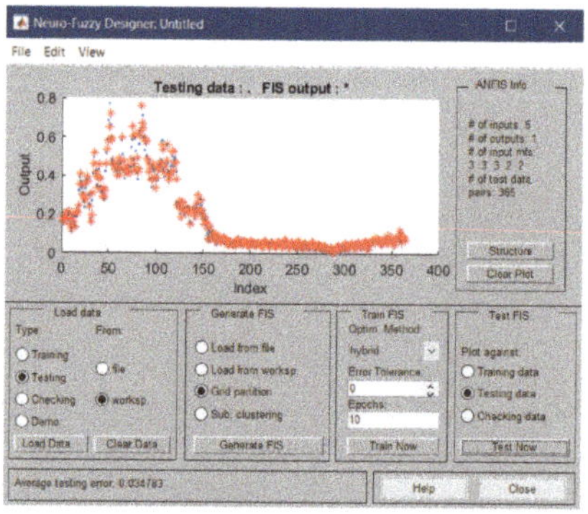

Figure 5. Screenshot of the testing data configuration.

In order to evaluate the prediction of NG demand, five well known and commonly used statistical indicators were introduced, i.e., mean square error (MSE), root mean square error (RMSE), mean absolute error (MAE), mean absolute percentage error (MAPE) and coefficient of determination (R^2). The mathematical equations of the statistical indicators are described below.

1. Mean squared error:

$$\mathrm{MSE} = \frac{1}{T}\sum_{t=1}^{T}(Z(t) - X(t))^2 \tag{9}$$

2. Root mean squared error:

$$\text{RMSE} = \sqrt{\text{MSE}} \tag{10}$$

3. Mean absolute error:

$$\text{MAE} = \frac{1}{T}\sum_{t=1}^{T}\left|Z(t) - X(t)\right| \tag{11}$$

4. Mean absolute percentage error:

$$\text{MAPE} = \frac{1}{T}\sum_{t=1}^{T}\left|\frac{Z(t) - X(t)}{Z(t)}\right| \tag{12}$$

5. Coefficient of determination:

$$\text{R} = \frac{T\sum_{t=1}^{T} Z(t)\cdot X(t) - \left(\sum_{t=1}^{T} Z(t)\right)\left(\sum_{t=1}^{T} X(t)\right)}{\sqrt{T\sum_{t=1}^{T}(Z(t))^2 - \left(\sum_{t=1}^{T} Z(t)\right)^2}\cdot\sqrt{T\sum_{t=1}^{T}(X(t))^2 - \left(\sum_{t=1}^{T} X(t)\right)^2}} \tag{13}$$

where $X(t)$ is the forecasted value of the NG at the t-th iteration, and $Z(t)$ is the actual value of the NG at the t-th iteration, $t = 1, \ldots, T$, where T is the number of testing records.

Higher values of R^2, i.e., closer to 1, mean better model performance and the regression line fits the data well. A coefficient of determination value of 1.0 points out that the regression curve fits the data perfectly.

3. Results

This section presents the exploration analysis results for the various ANFIS architectures as proposed in Section 2.2.2. Considering the steps proposed in Section 2.2, the initial dataset is split into training and testing. During the training process, the ANFIS model is designed for each one of the suggested configurations. After the training process of the ANFIS finishes, NG consumption demands for the next day (one day ahead prediction) are calculated from the generated FIS. The NG consumption results only for the city of Athens (as an indicative example) regarding all configurations tested, are presented in Table 4, whereas Table 5 gathers the best three results of NG consumption demand obtained from ANFIS for each of the 10 cities. To evaluate the performance of the models, the statistical indicator mean absolute percent error (MAPE) was used [91]. MAPE is a relative measurement, independent of scale, and it is the most common performance metric in time series forecasting, due to being reliable and valid [21].

Table 4. Testing results for the examined ANFIS architectures concerning the city of Athens.

Anfis Run	Type of Input MF	Number of MFs	Type of Output MF	Number of Epochs	Optimization	MSE	RMSE	MAE	MAPE	R^2
1	trimf	2-2-2-2-2	Constant	10	Hybrid	0.0010	0.0320	0.0192	12.6882	0.9849
2	trapmf	2-2-2-2-2	Constant	10	Hybrid	0.0013	0.0366	0.0245	19.8878	0.9806
3	gbellmf	2-2-2-2-2	Constant	10	Hybrid	0.0011	0.0335	0.0209	14.7498	0.9834
4	Gaussmf	2-2-2-2-2	Constant	10	Hybrid	0.0011	0.0326	0.0201	13.8422	0.9842
5	Gauss2mf	2-2-2-2-2	Constant	10	Hybrid	0.0011	0.0324	0.0197	13.5785	0.9845
6	pimf	2-2-2-2-2	Constant	10	Hybrid	0.0015	0.0389	0.0254	19.5486	0.9782
7	dsigmf	2-2-2-2-2	Constant	10	Hybrid	0.0014	0.0378	0.0244	18.7851	0.9794
8	psigmf	2-2-2-2-2	Constant	10	Hybrid	0.0014	0.0378	0.0244	18.7851	0.9794
9	trimf	2-2-3-3-3	Constant	10	Hybrid	0.0015	0.0388	0.0232	15.8840	0.9774
10	trapmf	2-2-3-3-3	Constant	10	Hybrid	0.0020	0.0448	0.0269	19.2727	0.9698
11	gbellmf	2-2-3-3-3	Constant	10	Hybrid	0.0014	0.0379	0.0227	15.5056	0.9785
12	Gaussmf	2-2-3-3-3	Constant	10	Hybrid	0.0014	0.0379	0.0226	15.7640	0.9784
13	Gauss2mf	2-2-3-3-3	Constant	10	Hybrid	0.0017	0.0410	0.0241	15.8227	0.9747
14	pimf	2-2-3-3-3	Constant	10	Hybrid	0.0130	0.1141	0.0347	21.6717	0.8552
15	dsigmf	2-2-3-3-3	Constant	10	Hybrid	0.0020	0.0448	0.0254	16.7809	0.9698

Table 4. *Cont.*

Anfis Run	Type of Input MF	Number of MFs	Type of Output MF	Number of Epochs	Optimization	MSE	RMSE	MAE	MAPE	R^2
16	psigmf	2-2-3-3-3	Constant	10	Hybrid	0.0020	0.0448	0.0254	16.7809	0.9698
17	trimf	3-3-3-2-2	Constant	10	Hybrid	0.0012	0.0348	0.0210	14.6116	0.9819
18	trapmf	3-3-3-2-2	Constant	10	Hybrid	0.0018	0.0430	0.0297	27.0255	0.9723
19	gbellmf	3-3-3-2-2	Constant	10	Hybrid	0.0013	0.0355	0.0212	14.4247	0.9810
20	Gaussmf	3-3-3-2-2	Constant	10	Hybrid	0.0011	0.0337	0.0198	12.8988	0.9829
21	trimf	3-3-3-3-3	Constant	10	hybrid	0.0021	0.0455	0.0242	15.4964	0.9698
22	trimf	3-3-3-3-3	Constant	10	backpropa	0.0559	0.2365	0.1610	74.9654	0.7447
23	trapmf	3-3-3-3-3	Constant	10	hybrid	0.0031	0.0556	0.0281	21.5814	0.9562
24	trapmf	3-3-3-3-3	Constant	10	backpropa	0.0501	0.2238	0.1527	72.2629	0.7404
25	gbellmf	3-3-3-3-3	Constant	10	hybrid	0.0014	0.0374	0.0217	14.6538	0.9791
26	gbellmf	3-3-3-3-3	Constant	10	backpropa	0.0015	0.0392	0.0265	25.1796	0.9793
27	trimf	3-3-3-3-3	Constant	30	hybrid	0.0016	0.0403	0.0224	13.3194	0.9759
28	trimf	3-3-3-3-3	Constant	50	hybrid	0.0017	0.0417	0.0224	13.2276	0.9745
29	trapmf	3-3-3-3-3	Constant	30	hybrid	0.0029	0.0539	0.0245	17.2238	0.9590
30	trapmf	3-3-3-3-3	Constant	50	hybrid	0.0017	0.0416	0.0233	16.4612	0.9745
31	gbellmf	3-3-3-3-3	Constant	30	hybrid	0.0013	0.0366	0.0213	13.2077	0.9799
32	gbellmf	3-3-3-3-3	Constant	50	hybrid	0.0019	0.0432	0.0236	13.4445	0.9724
33	trimf	3-3-4-4-4	Constant	10	hybrid	0.0023	0.0479	0.0251	15.4225	0.9662
34	trimf	3-3-5-5-5	Constant	10	hybrid	0.0078	0.0884	0.0320	17.3158	0.9006
35	trapmf	3-3-4-4-4	Constant	10	hybrid	0.0021	0.0454	0.0275	23.1769	0.9695
36	trapmf	3-3-5-5-5	Constant	10	hybrid	0.0098	0.1084	0.0450	19.3158	0.8806
37	gbellmf	3-3-4-4-4	Constant	10	hybrid	0.0022	0.0472	0.0256	16.1637	0.9669
38	gbellmf	3-3-5-5-5	Constant	10	hybrid	0.0044	0.0660	0.0307	18.0977	0.9376
39	gaussmf	3-3-3-3-3	Constant	10	hybrid	0.0013	0.0365	0.0212	13.8235	0.9800
40	gaussmf	3-3-4-4-4	Constant	10	hybrid	0.0019	0.0431	0.0241	14.7715	0.9720
41	gaussmf	3-3-5-5-5	Constant	10	hybrid	0.0056	0.0746	0.0314	17.5307	0.9185
42	gauss2mf	3-3-3-3-3	Constant	10	hybrid	0.0017	0.0409	0.0235	16.2626	0.9755
43	gauss2mf	3-3-4-4-4	Constant	10	hybrid	0.0040	0.0632	0.0260	17.3863	0.9407
44	gauss2mf	3-3-5-5-5	Constant	10	hybrid	0.0072	0.0847	0.0290	17.7331	0.9048
45	pimf	3-3-3-3-3	Constant	10	hybrid	0.1224	0.3499	0.0482	26.0901	0.3608
46	pimf	3-3-4-4-4	Constant	10	hybrid	0.0026	0.0510	0.0307	24.7626	0.9615
47	pimf	3-3-5-5-5	Constant	10	hybrid	0.0022	0.0466	0.0285	22.3553	0.9678

Table 5. Testing results for the best three ANFIS architectures of each city based on MAPE value.

City	Anfis Run	Type of Input MF	Number of MFs	Number of Rules	Time (s)	MSE	RMSE	MAE	MAPE	R^2
Alexandroupoli	17	trimf	3-3-3-2-2	72	5	0.0024	0.0494	0.0351	10.5278	0.9638
	39	gaussmf	3-3-3-3-3	243	47	0.0031	0.0557	0.0355	10.1556	0.9538
	20	gaussmf	3-3-3-2-2	72	5	0.0023	0.0480	0.0341	10.1123	0.9659
Athens	1	trimf	2-2-2-2-2	32	7	0.0021	0.0457	0.0295	20.1799	0.9825
	17	trimf	3-3-3-2-2	108	19	0.0026	0.0511	0.0315	19.7972	0.9786
	20	gaussmf	3-3-3-2-2	108	19	0.0022	0.0467	0.0306	21.2929	0.9818
Drama	17	trimf	3-3-3-2-2	108	19	0.0026	0.0511	0.0363	6.2547	0.8997
	1	trimf	2-2-2-2-2	32	5	0.0026	0.0513	0.0361	6.2235	0.8975
	20	gaussmf	3-3-3-2-2	108	13	0.0026	0.0508	0.0371	6.4071	0.8995
Karditsa	17	trimf	3-3-3-2-2	108	12	0.0019	0.0434	0.0242	13.8394	0.9789
	1	trimf	2-2-2-2-2	32	4	0.0018	0.0421	0.0236	11.6196	0.9801
	4	gaussmf	2-2-2-2-2	32	4	0.0019	0.0431	0.0248	13.3841	0.9792
Larissa	1	trimf	2-2-2-2-2	32	4	0.0012	0.0352	0.0203	10.9568	0.9817
	4	gaussmf	2-2-2-2-2	32	4	0.0012	0.0352	0.0204	10.9833	0.9817
	20	gaussmf	3-3-3-2-2	108	19	0.0010	0.0314	0.0184	10.5236	0.9858
Markopoulo	1	trimf	2-2-2-2-2	32	5	0.0091	0.0956	0.0728	25.0887	0.6593
	4	gaussmf	2-2-2-2-2	32	5	0.0096	0.0980	0.0755	26.7510	0.6364
	17	trimf	3-3-3-2-2	108	19	0.0259	0.1609	0.1087	36.7174	0.5126
Serres	1	trimf	2-2-2-2-2	32	5	0.0007	0.0271	0.0176	10.4721	0.9839
	4	gaussmf	2-2-2-2-2	32	5	0.0008	0.0279	0.0185	11.2421	0.9831
	39	gaussmf	3-3-3-3-3	243	45	0.0008	0.0285	0.0194	12.1163	0.9824
Thessaloniki	17	trimf	3-3-3-2-2	108	13	0.0015	0.0382	0.0229	16.1046	0.9773
	20	gaussmf	3-3-3-2-2	108	13	0.0013	0.0363	0.0219	14.1944	0.9795
	39	gaussmf	3-3-3-3-3	243	45	0.0021	0.0459	0.0256	15.2032	0.9672
Trikala	1	trimf	2-2-2-2-2	32	4	0.0019	0.0433	0.0232	10.5817	0.9815
	4	gaussmf	2-2-2-2-2	32	4	0.0020	0.0450	0.0245	11.1412	0.9800
	20	gaussmf	3-3-3-2-2	108	13	0.0028	0.0530	0.0271	11.7631	0.9708
Volos	1	trimf	2-2-2-2-2	32	4	0.0021	0.0459	0.0317	13.2520	0.9564
	4	gaussmf	2-2-2-2-2	32	4	0.0021	0.0460	0.0314	13.1629	0.9563
	20	gaussmf	3-3-3-2-2	108	12	0.0020	0.0445	0.0323	13.9710	0.9588

The corresponding graphical representation of the results regarding the best three out of 47 total ANFIS architectures for the city of Athens is illustrated in Figure 6. Also, the best ANFIS model for each city can be found in Table 6, which provides the most reliable ANFIS architecture results for each city. All the results have been previously ranked, based on the minimum value of MAPE and subsequently the minimum values of MSE, RMSE and MAE. The priority was given to MAPE as one of the most crucial evaluation metrics, according to the literature [91,96], which was used in this study to compare various models obtained from ANFIS and other soft computing and neural networks methods. As a relative and easy to interpret measurement, MAPE is reliable, valid and independent of scale. The smaller the values of MAPE are, the closer the forecasted values are to the actual values.

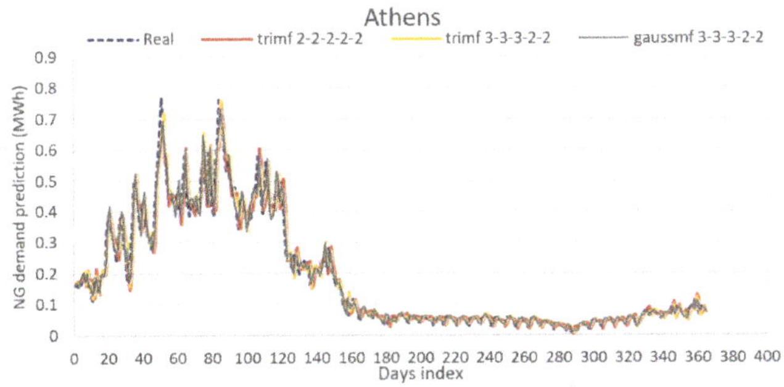

Figure 6. Forecasting results for the best three ANFIS architectures for Athens.

Table 6. The best ANFIS models for all the cities under investigation (10 epochs and hybrid optimization).

Title 1	Anfis Run	Type of Input MF	Number of MFs	Type of Output MF	Optimization	MSE	RMSE	MAE	MAPE	R^2
Alexandroupoli	20	gaussmf	3-3-3-2-2	Constant	Hybrid	0.0023	0.0480	0.0341	10.1123	0.9659
Athens	17	trimf	3-3-3-2-2	Constant	Hybrid	0.0026	0.0511	0.0315	12.7972	0.9706
Drama	1	trimf	2-2-2-2-2	Constant	Hybrid	0.0026	0.0513	0.0361	6.2235	0.8975
Karditsa	1	trimf	2-2-2-2-2	Constant	Hybrid	0.0018	0.0421	0.0236	11.6196	0.9801
Larissa	20	gaussmf	3-3-3-2-2	Constant	Hybrid	0.0010	0.0314	0.0184	10.5236	0.9858
Markopoulo	1	trimf	2-2-2-2-2	Constant	Hybrid	0.0091	0.0956	0.0728	25.0887	0.6593
Serres	4	gaussmf	2-2-2-2-2	Constant	Hybrid	0.0008	0.0279	0.0185	11.2421	0.9831
Thessaloniki	20	gaussmf	3-3-3-2-2	Constant	Hybrid	0.0013	0.0363	0.0219	14.1944	0.9795
Trikala	4	gaussmf	2-2-2-2-2	Constant	Hybrid	0.0020	0.0450	0.0245	11.1412	0.9800
Volos	4	gaussmf	2-2-2-2-2	Constant	Hybrid	0.0021	0.0460	0.0314	13.1629	0.9563

As illustrated in Table 6, ANFIS models appear to perform best mostly when triangular MFs are used for the input variables: three MFs for the first three input variables (month, day of week and mean temperature) and two or three MFs for the other two input variables (daily demand for current day and one day before). Also, constant MFs are selected for the output variable and hybrid optimization method. The graphical representation of the best ANFIS models for each city is illustrated in Figure 7.

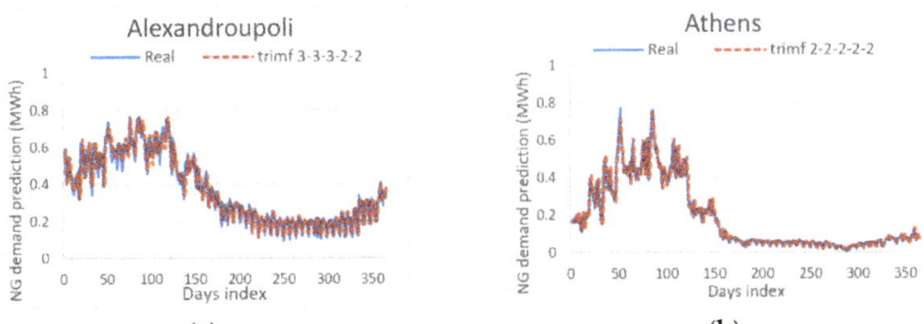

(a) (b)

Figure 7. *Cont.*

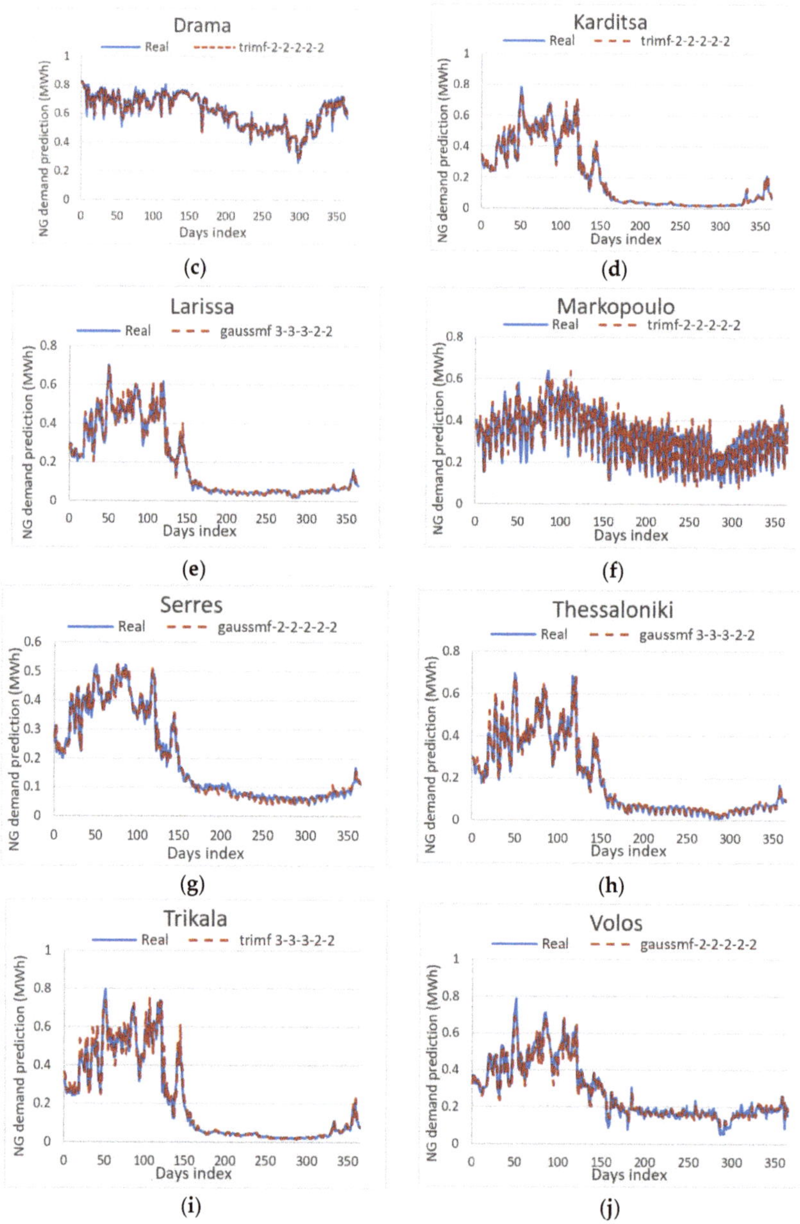

Figure 7. Forecasting results for three cities considering the best ANFIS method. (**a**) Testing Alexandroupoli, (**b**) testing Athens, (**c**) testing Drama, (**d**) testing Karditsa, (**e**) testing Larissa, (**f**) testing Markopoulo, (**g**) testing Serres, (**h**) testing Thessaloniki, (**i**) testing Trikala, (**j**) testing Volos.

It is worth mentioning that all three most efficient ANFIS architectures with respect to MAPE values have triangular or gaussian MFs and 2-2-2-2-2 or 3-3-3-2-2 number of input MFs, whereas the output MF is constant and the learning algorithm is hybrid. In addition, the application of other MFs combinations does not seem to give results that could be on top of the list. Due to the limitation of the total number of parameters that should not exceed the number of training data pairs, the number of MFs was chosen based on the number of input parameters. Figure 8 shows the exponential increase in the number of rules when the number of MFs increases, whereas Table A2 in Appendix A gathers the time and number of rules for all the proposed ANFIS configurations.

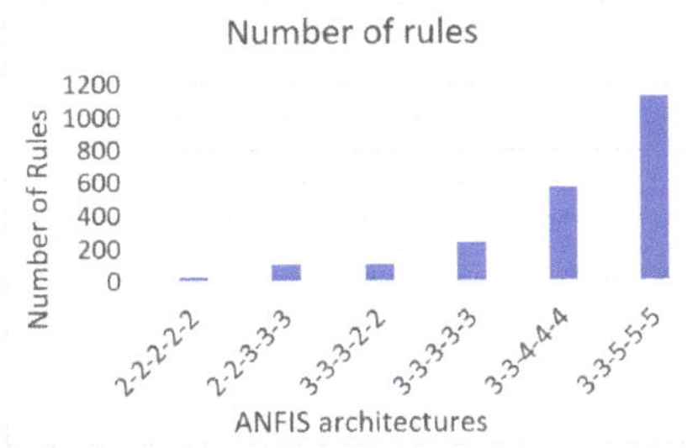

Figure 8. Number of rules for various ANFIS architectures.

3.1. Comparison with ANNs, FCMs and Hybrid FCM-ANN

To further investigate the performance of the proposed ANFIS architectures, an extensive comparative analysis between the state of the art ANNs, soft computing methods of FCMs and their hybrid combination of FCMs with ANNs was performed.

The architecture of the analyzed ANN was a multilayer feed forward network with an input layer containing five inputs (month, day, temperature, demand of a day before, current demand), a hidden layer with 10 neurons, and an output layer with one output (a day-ahead demand prediction). The authors used the sigmoidal activation function in all layers and implemented Levenberg–Marquardt algorithm to train the network.

The soft computing method of fuzzy cognitive map with evolutionary learning capabilities, such as the real-coded genetic algorithm (RCGA-FCM) and structure optimization genetic algorithm (SOGA-FCM) [100], were used for time series modeling and prediction of day-ahead NG energy demand. For FCM learning, we implemented RCGA-FCM and SOGA-FCM. A short description on the applied evolutionary-based FCM approaches is given in Appendix B. The implementations of FCMs differ from ANFIS, even though they both belong to the soft computing family.

In this research study, we used the dynamic model type (Equation (B1)) which is found in Appendix B, with sigmoidal transformation function. FCMs learned with the use of RCGA and SOGA algorithm contain five concepts (month, day, temperature, demand of a day before, current demand) [114].

The applied hybrid approach for time series prediction is based on FCMs and ANNs and was previously proposed in [18,19]. It allows us to select the most significant concepts for FCM using SOGA. These concepts are used as the inputs for ANN. In the hybrid approach, we used artificial neural networks with an input layer with five inputs selected by the SOGA-FCM approach, a hidden layer with 10 neurons and an output layer with one output (one day-ahead demand prediction). Sigmoidal activation function and Levenberg–Marquardt learning algorithm were used. All the simulations for FCMs and hybrid FCM-ANN configurations were performed with the software tool ISEMK [115] which has been developed for time series forecasting purposes. An analytical description of FCM-based models and hybrid FCM-ANN can be found in [13,101,102,116], whereas they are used in this work only for comparison purposes.

In what follows, Table 7 gathers the results of the explored ANN and soft computing models, which are straightforward compared with our best performed ANFIS configuration, for each one out of the 10 cities, suggested in this research work. In Figure 9, three indicative graphs of the cities Alexandroupoli, Athens, and Drama are illustrated regarding the predicted values of NG demand for all the best proposed architectures. In Appendix A, the corresponding graphs of the rest of the cities are presented (see Figure A1).

Table 7. Comparison results among the artificial neural network (ANN), fuzzy cognitive map (FCM), hybrid FCM-ANN and best ANFIS architectures of each city.

City	Method	MSE	RMSE	MAE	MAPE	R^2
Alexandroupoli	RCGA-FCM	0.0047	0.0684	0.0538	17.6233	0.9450
	SOGA-FCM	0.0045	0.0672	0.0526	17.1707	0.9484
	ANN	0.0042	0.0645	0.0505	16.1131	0.9439
	Hybrid FCM-ANN	0.0034	0.0579	0.0427	14.3034	0.9498
	Best ANFIS	0.0023	0.0480	0.0341	10.1123	0.9659
Athens	RCGA-FCM	0.0022	0.0473	0.0303	23.5985	0.9676
	SOGA-FCM	0.0029	0.0539	0.0337	22.7453	0.9646
	ANN	0.0010	0.0323	0.0198	14.2464	0.9844
	Hybrid FCM-ANN	0.0014	0.0374	0.0230	17.5418	0.9790
	Best ANFIS	0.0026	0.0511	0.0315	19.7972	0.9786
Drama	RCGA-FCM	0.0080	0.0894	0.0749	12.9942	0.8691
	SOGA-FCM	0.0056	0.0748	0.0600	10.1766	0.8796
	ANN	0.0025	0.0501	0.0357	6.1657	0.9025
	Hybrid FCM-ANN	0.0028	0.0526	0.0363	6.2502	0.8941
	Best ANFIS	0.0026	0.0513	0.0361	6.2235	0.8975
Karditsa	RCGA-FCM	0.0039	0.0624	0.0379	27.5914	0.9591
	SOGA-FCM	0.0488	0.2210	0.1397	50.2112	0.9711
	ANN	0.0016	0.0405	0.0245	17.4579	0.9819
	Hybrid FCM-ANN	0.0017	0.0407	0.0245	18.4095	0.9817
	Best ANFIS	0.0018	0.0421	0.0236	11.6196	0.9801
Larissa	RCGA-FCM	0.0027	0.0515	0.0331	22.2481	0.9638
	SOGA-FCM	0.0025	0.0505	0.0328	22.9579	0.9649
	ANN	0.0013	0.0355	0.0209	13.2479	0.9812
	Hybrid FCM-ANN	0.0013	0.0356	0.0215	13.1974	0.9811
	Best ANFIS	0.0010	0.0314	0.0184	10.5236	0.9858
Markopoulo	RCGA-FCM	0.0075	0.0868	0.0726	26.0003	0.6975
	SOGA-FCM	0.0078	0.0883	0.0739	26.3345	0.6955
	ANN	0.0172	0.1310	0.1048	34.8594	0.4765
	Hybrid FCM-ANN	0.0070	0.0836	0.0667	23.7166	0.7094
	Best ANFIS	0.0091	0.0956	0.0728	25.0887	0.6593
Serres	RCGA-FCM	0.0017	0.0409	0.0274	16.5199	0.9648
	SOGA-FCM	0.0495	0.2225	0.1632	72.9785	0.9772
	ANN	0.0008	0.0275	0.0179	10.9948	0.9842
	Hybrid FCM-ANN	0.0008	0.0289	0.0190	11.5000	0.9821
	Best ANFIS	0.0008	0.0279	0.0185	11.2421	0.9831
Thessaloniki	RCGA-FCM	0.0029	0.0541	0.0339	29.9713	0.9565
	SOGA-FCM	0.0029	0.0539	0.0340	30.1471	0.9568
	ANN	0.0017	0.0412	0.0262	23.8748	0.9735
	Hybrid FCM-ANN	0.0019	0.0441	0.0266	23.8835	0.9696
	Best ANFIS	0.0013	0.0363	0.0219	14.1944	0.9795
Trikala	RCGA-FCM	0.0059	0.0770	0.0453	21.9722	0.9528
	SOGA-FCM	0.0433	0.2082	0.1287	42.7427	0.9715
	ANN	0.0020	0.0443	0.0258	14.1183	0.9804
	Hybrid FCM-ANN	0.0019	0.0432	0.0251	13.9034	0.9815
	Best ANFIS	0.0020	0.0450	0.0245	11.1412	0.9800
Volos	RCGA-FCM	0.0028	0.0526	0.0397	17.8195	0.9436
	SOGA-FCM	0.0027	0.0520	0.0395	17.8988	0.9445
	ANN	0.0020	0.0444	0.0319	13.2504	0.9588
	Hybrid FCM-ANN	0.0020	0.0446	0.0307	12.7881	0.9587
	Best ANFIS	0.0021	0.0460	0.0314	13.1629	0.9563

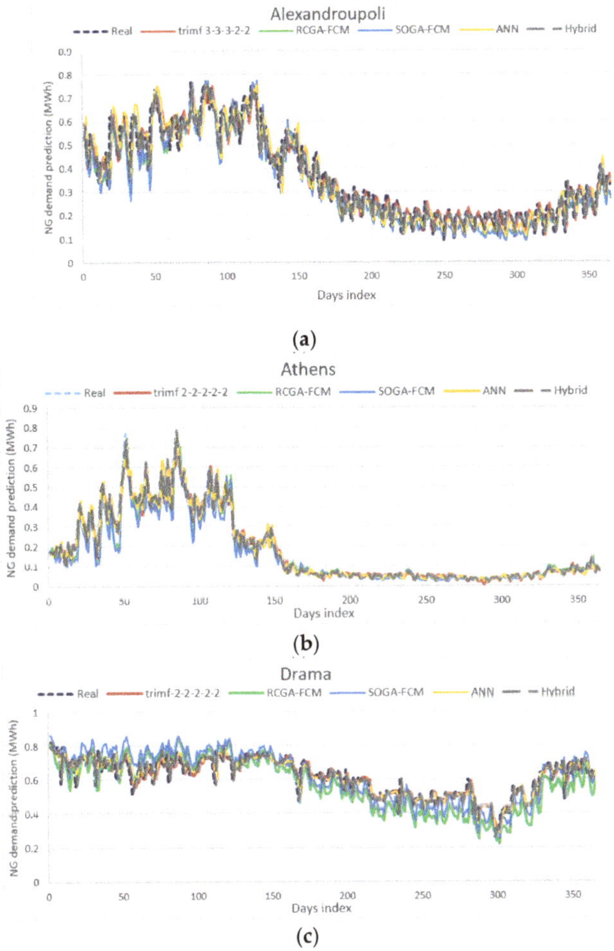

Figure 9. Comparison of forecasting results for each city considering all examined methods. (**a**) Testing for Alexandroupoli, (**b**) Testing for Athens, (**c**) Testing for Drama.

For a deeper analysis of the examined architectures (ANFIS, ANN, FCM, hybrid FCM-ANN), the authors report on further details regarding the parameters of each model used in this study. The ANN and FCM models were previously applied for NG demand prediction in several research works, such as those in [13,101,102,116]. The models were sufficiently described and the hyperparameters were properly configured to offer optimum performance of the investigated FCM models.

Table 8 depicts the optimum parameters for all cities considering the neural and FCM evolutionary methods (ANN, RCGA-FCM, SOGA-FCM, Hybrid), compared with the proposed best performed ANFIS. The average running time is also presented in Table 8, which was calculated for each soft computing architecture for all models. It is worth mentioning that we have conducted a rigorous exploratory analysis for all the investigated neuro-fuzzy, soft computing techniques and ANNs, with different parameters, for training and model optimization, to reach the highest prediction accuracy with respect to the evaluation metrics.

Table 8. Parameters and average running time for each architecture.

Architectures	Parameters for Athens City	Average Running Time
ANN	Multilayer feed forward network, six inputs, 10 neurons, one output, sigmoidal activation function, Levenberg-Marquardt learning, epochs = 20	16–20 s
RCGA-FCM	Uniform crossover with probability 0.4, Mühlenbein's mutation with probability 0.4, ranking selection, elite strategy, population size 200, maximum number of generations 200	808 s
SOGA-FCM	Uniform crossover with probability 0.4, Mühlenbein's mutation with probability 0.4, ranking selection, elite strategy, population size 200, maximum number of generations 200, learning parameters $b_1 = b_2 = 0.01$	799 s
Hybrid FCM-ANN	Multilayer feed forward network, four inputs selected by SOGA-FCM (month, temperature, demand of a day before, current demand), one hidden layer with 10 neurons, one output, sigmoidal activation function, Levenberg-Marquardt learning, epochs = 20	811 s
Best ANFIS	Triangular mf, 2-2-2-2-2 or 3-3-3-2-2, Constant output, epochs = 10, Hybrid optimization	4–19 s

3.2. Discussion of Results

In this work, several ANFIS architectures were investigated, with respect to all the variables that were carefully determined in the developed model as reported in Section 2.1 and after different sets of model configurations were tested. However, only one ANFIS architecture reached the optimum performance, in terms of forecasting accuracy, considering the minimum value of MAPE and subsequently the minimum values of MSE, RMSE and MAE values produced. In particular, it emerged that the optimum ANFIS configuration is 2-2-2-2-2 with triangular MFs for input variables, which produces the most simple (concerning the number of rules), fast (see Table A2 in Appendix A) and accurate model for this energy forecasting problem. In general, it is observed that the best results are produced from the combination of triangular or gaussian MFs regarding the input variables, and the constant MFs regarding the output layers.

To further discuss the results produced and to show the effectiveness of the proposed forecasting methodology of ANFIS, the authors conducted a comparative analysis regarding the forecasting performance between the proposed technique, and other ANN and soft computing methods too, such as FCM, which were reported in the literature and have already been applied in the specific domain. The MAPE criterion was used to compare various models from ANN, evolutionary FCM, Hybrid FCM-ANN and ANFIS. The results are given in Table 7. For example, the MAPE values of RCGA-FCM, SOGA-FCM, ANN, hybrid and ANFIS models for the city of Alexandroupoli were calculated as 17.62%, 17.17%, 16.11%, 14.30% and 10.11%, respectively. The smaller the values of MAPE are, the closer to the actual values the forecasted values are. The best result was obtained from the ANFIS model. The respective figures in the text and in Appendix A have been updated with the new prediction values.

Considering the same dataset linked to only three cities (Athens, Thessaloniki, and Larissa) out of the ten that participated in our study, a day-ahead NG consumption prediction was investigated in [117], applying ANN and LSTM approaches and in [102], implementing the SOGA-FCM method and a hybrid combination of it. Furthermore, an ensemble FCM prediction methodology concerning the same dataset was presented in [13], in which a recent soft computing technique for time series forecasting, using evolutionary fuzzy cognitive maps and their ensemble combination was compared to ANNs, as benchmark forecasting methods. These methods and their results in terms of MSE and MAE values for three benchmark cities are all gathered in the following table and certain results can be concluded. The main reason for selecting the statistical indicators MSE and MAE in the following figure is to accomplish a straightforward comparison with the results published in previous works.

In Figure 10, it can be noted that all methods achieve high accuracy in NG consumption predictions, using the same dataset. The best ANFIS approach seems to excel over the ensemble and hybrid methods. Consequently, the proposed ANFIS architecture, which handles the fuzziness of data more efficiently, outperforms all the other examined methods in most cases, with a rather remarkable difference. ANFIS is less time consuming and more flexible than ANN, and as it employs fuzzy rules and membership functions incorporating with real-world systems, it can be used as alternate method to ANN forecasting.

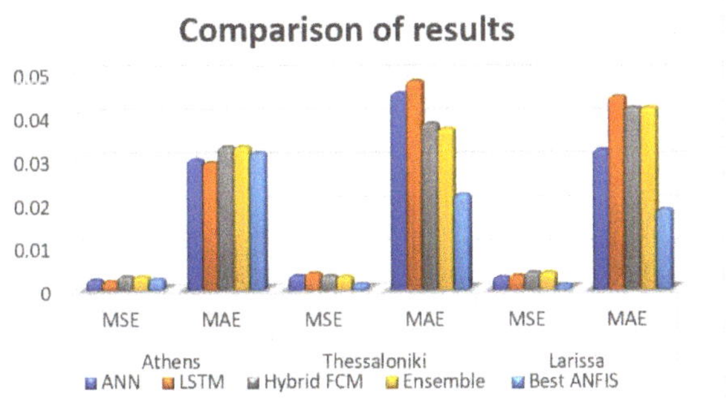

Figure 10. Comparison of results between machine learning and soft computing methods for three benchmark cities.

It is observed that the proposed method exhibits better or similar performance to other well-known ANN, FCM or hybrid FCM-ANN architectures for the ten cities under investigation. The produced results highlight the significance and superiority of neuro-fuzzy methods over the other examined methods in terms of prediction accuracy, when they deal with time series forecasting problems in energy. This is in accordance with the main advantageous features of ANFIS models, which are their ability to capture the nonlinear structure of a process, their adaptation capability, and fast training characteristics. As reported in the literature, ANFIS models are able to cope with the uncertainty and fuzziness that characterize the energy domain [118,119] when other intelligent methods cannot tackle them.

The main outcomes of this study can be summarized as follows:

i. The proposed ANFIS method exhibits the best performance when certain configuration settings are selected for the examined datasets which are linked to ten cities of Greece. The authors concluded that a certain configuration is best for the examined ANFIS model, after having conducted a number of experiments and following a trial-and error approach. The best ANFIS model is based on a distinct architecture that features a 2-2-2-2-2 triangular or gaussian MF.
ii. The proposed ANFIS architecture is superior to the four benchmark and well-known ANN and FCM methods (ANN, SOGA-FCM, RCGA-FCM, Hybrid FCM-ANN), which have been efficiently used in NG consumption forecasting. The results presented in Table 7, which gathers various error indicators and the R^2, as prediction accuracy indices for all five architectures, show that the best ANFIS model holds the best prediction accuracy among all the methods that were included in this comparative analysis.
iii. The proposed ANFIS model shows significant capacity when applied to forecasting NG demand, since it exhibits better performance (see Table 7) with less running time (see Table 8) and more flexibility to handle fuzziness than other well-known ANN and FCM architectures.

4. Conclusions

This study proposes the ANFIS method to predict short-term demand of NG consumption. This approach is applied on the Greek territory and uses 10 different datasets provided by DESFA, that regard previous energy consumption historical data, for ten main cities. To decide the model's proper architecture, the authors follow an exploration process regarding the best configuration of input and training parameters. The best ANFIS model is then compared to other well-known ANN and soft computing models that are commonly used for energy demand prediction purposes. The ANFIS method demonstrates significant performance in the field of energy demand prediction, outweighing the traditional ANN and FCM architectures. In addition, the running time of the proposed architecture is much less than those of other examined models, making it the right decision for day-ahead demand forecasting of NG. The findings of this study reveal that the highest forecasting accuracy emerged when the same model configuration was used for most of the cities, highlighting the generalization capabilities of the proposed architecture.

This work can be widely used in short-term demand forecasting for other countries too, with the same or similar input parameters, and can be also useful especially for distribution operators, providing them with the ability to make long-term planning decisions and apply the correct strategic policies in this direction. Following the literature, the ANFIS approach can be applied in various other domains such as medicine, environmental modelling, various energy systems, like solar and wind, as well as other engineering applications. As can be seen, the ANFIS application area is wide, and as regards the energy sector, this method finds great applicability due to its high prediction accuracy, robustness, and easiness to use.

The results show that the proposed algorithm, which was proven to be efficient, fast and robust, can be adopted by regulatory authorities and decision makers to perform rigorous forecasting of natural gas demand for the respective case cities and other cities in Greece too. The investigated approach is an accurate estimation method as it makes efficient short-term predictions in natural gas demand, showing minor deviations between the real and the predicting values. Since short-term natural gas forecasting is mostly used for the timely reservation of transport, storage capacity optimization, timely purchase of natural gas deliveries and capacity allocation, this method becomes critical to determine the energy policy for Greece and the wider area too, having overall a positive impact in natural gas consumption.

Future work is oriented in developing more advanced neuro-fuzzy models providing explainability and transparency in prediction tasks in diverse research domains, in order to evaluate the generalization capabilities of this approach. Furthermore, new forecast combination architectures of efficient deep learning and regularized recurrent neural networks for time series modelling and prediction in the energy sector will be investigated.

Author Contributions: Conceptualization, E.I.P. and E.I.P.; methodology, E.I.P.; software, E.I.P., K.P. (Konstantinos Papageorgiou) and K.P. (Katarzyna Poczeta); validation, E.I.P., E.I.P., K.P. (Konstantinos Papageorgiou), K.P. (Katarzyna Poczeta), D.B. and G.S.; formal analysis, E.I.P. and E.I.P.; investigation, E.I.P.; resources, E.I.P.; data curation, E.I.P. and K.P. (Konstantinos Papageorgiou), K.P. (Katarzyna Poczeta); writing—original draft preparation, E.I.P.; writing—review and editing, K.P. (Katarzyna Poczeta), E.I.P., D.B. and G.S.; visualization, E.I.P.; supervision, E.I.P. and G.S.; project administration, E.I.P. and D.B. All authors have read and agreed to the published version of the manuscript.

Funding: This research received no external funding.

Conflicts of Interest: The authors declare no conflict of interest.

Appendix A

Table A1. Different configurations of the selected ANFIS architectures regarding linear output MF.

Type of Input MF	Number of MFs	Type of Output MF	Number of Rules	MSE	RMSE	MAE	MAPE	R^2	Time (s)
trimf	2-2-2-2-2	Linear	32	0.001195	0.034572	0.019426	11.72180	0.982121	148
trapmf	2-2-2-2-2	Linear	32	0.001358	0.036859	0.020861	12.12378	0.979559	148
gbellmf	2-2-2-2-2	Linear	32	0.001267	0.035603	0.019921	11.46446	0.980963	148
Gaussmf	2-2-2-2-2	Linear	32	0.001298	0.036038	0.020259	11.97794	0.980468	148
Gauss2mf	2-2-2-2-2	Linear	32	0.001406	0.037496	0.020878	11.26382	0.978860	148
pimf	2-2-2-2-2	Linear	32	0.001635	0.040442	0.022176	12.08298	0.975405	148
dsigmf	2-2-2-2-2	Linear	32	0.001423	0.037733	0.021062	11.26721	0.978592	148
psigmf	2-2-2-2-2	Linear	32	0.001423	0.037733	0.021062	11.26722	0.978592	148
trimf	2-2-3-3-3	Linear	108	0.001476	0.038430	0.020941	11.17862	0.977773	328
Gaussmf	2-2-3-3-3	Linear	108	0.002038	0.045149	0.023286	12.71720	0.969241	328

Table A2. Running time and number of rules for all the proposed ANFIS configurations.

Type of Input MF	Number of MFs	Type of Output MF	Number of Epochs	Optimization	Number of Rules	Time Run
trimf, trapmf, gbell, gauss, pim, sigm	2-2-2-2-2	Constant	10	Hybrid	32	7 s
trimf, trapmf, gbell, gauss, pim, sigm	2-2-3-3-3	Constant	10	Hybrid	108	11 s
trimf, trapmf, gbell, gauss, pim, sigm	3-3-3-2-2	Constant	10	Hybrid	108	19 s
trimf, trapmf, gbell	3-3-3-3-3	Constant	10	Hybrid	243	68 s
trimf	3-3-4-4-4	Constant	10	Hybrid	576	10 min 10 s
trimf	3-3-5-5-5	Constant	10	Hybrid	1125	40 min
trapmf	3-3-4-4-4	Constant	10	Hybrid	576	12min
trapmf	3-3-5-5-5	Constant	10	Hybrid	1125	70 min
gbellmf	3-3-4-4-4	Constant	10	Hybrid	576	12 min 35 s
gbellmf	3-3-5-5-5	Constant	10	Hybrid	1125	50 min
gaussmf	3-3-3-3-3	Constant	10	Hybrid	243	4 min
gaussmf	3-3-4-4-4	Constant	10	Hybrid	576	25 min
gaussmf	3-3-5-5-5	Constant	10	Hybrid	1125	47 min
gauss2mf	3-3-3-3-3	Constant	10	Hybrid	243	4 min
gauss2mf	3-3-4-4-4	Constant	10	Hybrid	576	25 min
gauss2mf	3-3-5-5-5	Constant	10	Hybrid	1125	47 min
pimf	3-3-3-3-3	Constant	10	Hybrid	243	3.5 min
pimf	3-3-4-4-4	Constant	10	hybrid	576	20 min
pimf	3-3-5-5-5	Constant	10	hybrid	1125	42 min

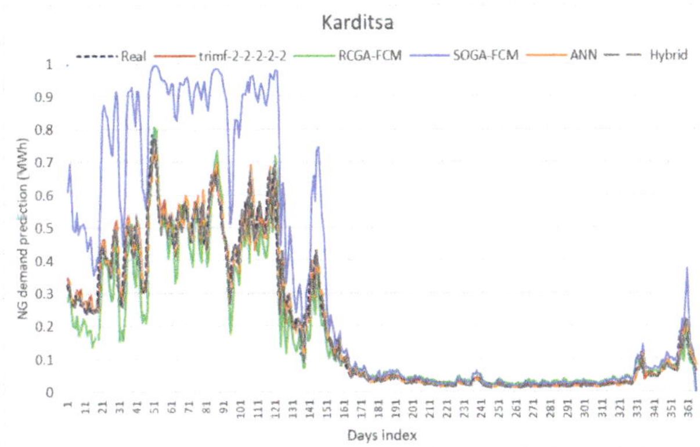

(a)

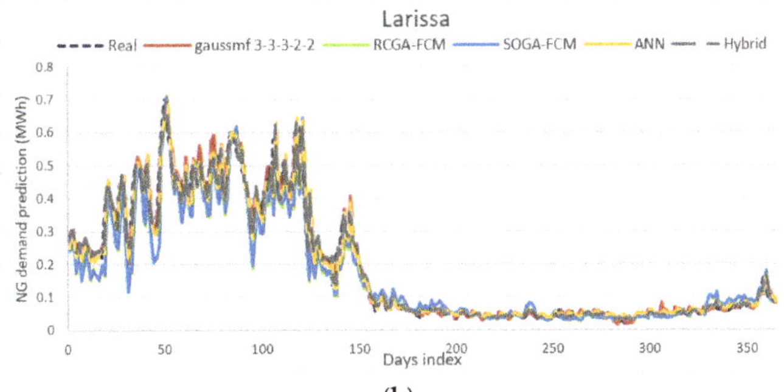

(b)

Figure A1. *Cont.*

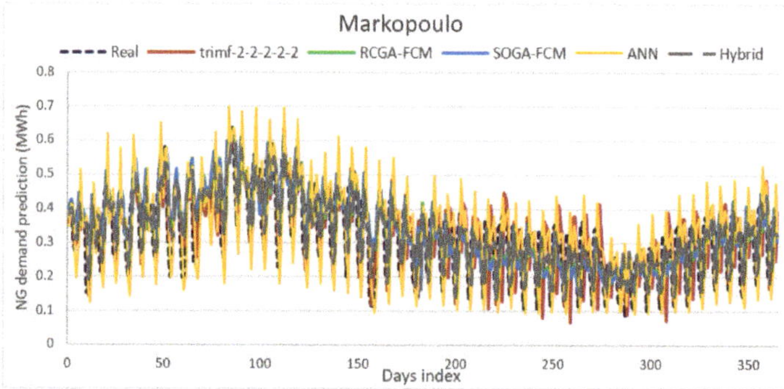

(c)

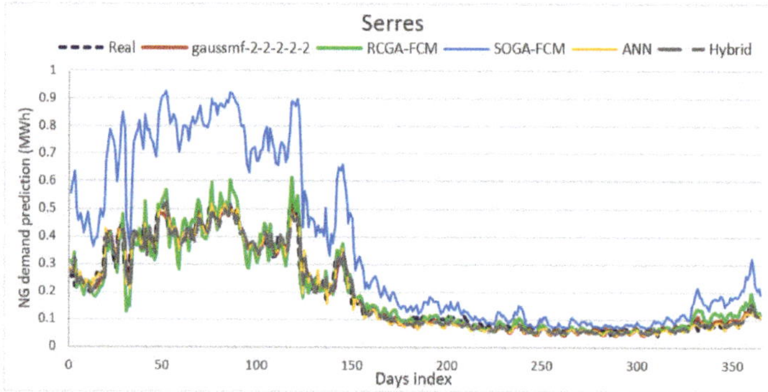

(d)

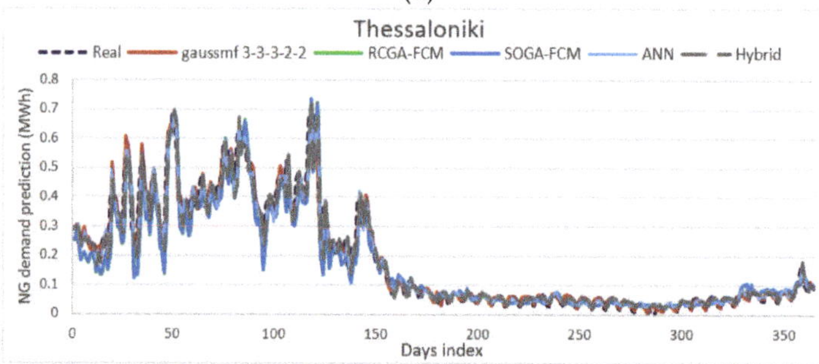

(e)

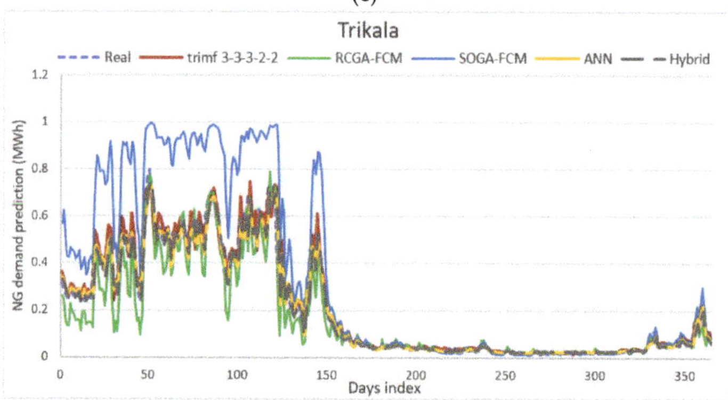

(f)

Figure A1. *Cont.*

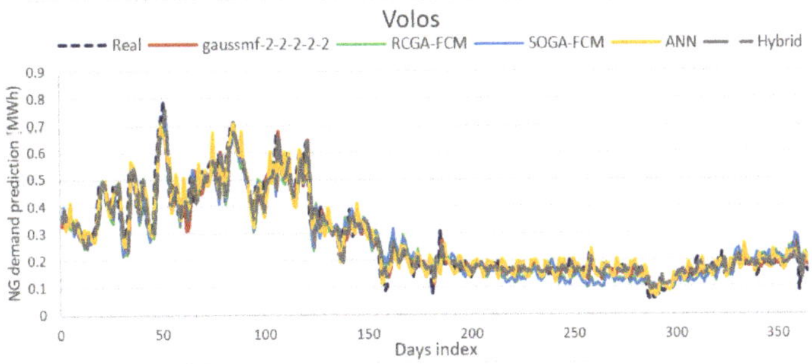

(g)

Figure A1. Comparison of forecasting results for each city considering all examined methods. (**a**) Testing for Karditsa, (**b**) testing for Larissa, (**c**) testing for Markopoulo, (**d**) testing for Serres, (**e**) testing for Thessaloniki, (**f**) testing for Trikala, (**g**) testing for Volos.

Appendix B

Appendix B.1. Fuzzy Cognitive Maps

Fuzzy cognitive maps (FCMs) are an effective tool for modeling and predicting time series. The structure of the FCM model is based on a directed graph, the nodes of which denote concepts significant for the analyzed problem, and the links are the causal relationships. Values of concepts can change over time according to the adopted dynamics model, for example the nonlinear dynamics model:

$$X_i(t+1) = F\left(X_i(t) + \sum_{\substack{j=1 \\ j \neq i}}^{n} X_j(t) \cdot w_{j,i}\right) \tag{A1}$$

where $X_i(t)$ is the value of the i-th concept at the t-th iteration, $w_{j,i}$ is the weight of the causal relationship between concepts X_j and X_i taking values from the range [−1,1], t is discrete time, $i,j = 1, 2, \ldots, n$, n is the number of concepts, and F is the transformation function normalizing the factor values to the range [0,1] or [−1,1]. Fuzzy cognitive maps can be constructed based on expert knowledge or with the use of machine learning algorithms. The aim of fuzzy cognitive map learning is to determine the weights of the causal relationships between concepts on the basis of available time series.

An effective method for fuzzy cognitive map learning is the real-coded genetic algorithm (RCGA) [100]. RCGA defines each individual in the population based on a floating-point vector containing the causal relationships. Each individual is decoded into a candidate map and evaluated with the use of proper fitness function. We used the following fitness function:

$$fitness_p(erorr_l) = \frac{1}{a \cdot erorr_l + 1} \tag{A2}$$

where a is a parameter, l is the number of generations, $l = 1, \ldots, L$, L is the maximum number of generations, p is the number of individuals, $p = 1, \ldots, P$, P is the population size, and $erorr_l$ is the learning error that can be in the following form:

$$erorr_l = \frac{1}{T}\sum_{t=1}^{T}(Z(t) - X(t))^2 \tag{A3}$$

where $X(t)$ is the predicted value of the decision concept at the t-th iteration, $Z(t)$ is the real normalized value of the decision concept at the t-th iteration, $t = 1, \ldots, T$, and T is the number of learning records.

Another way of learning fuzzy cognitive maps is the structure optimization genetic algorithm (SOGA) [100,120]. This allows one to simplify the structure of the FCM model by selecting the most significant concepts and causal relationships during the learning process. In this approach, the fitness function is based on the modified learning error including an additional penalty for highly complexity of the candidate fuzzy cognitive map, understood as a large number of concepts and non-zero relationships, described as follows:

$$erorr'_l = erorr_l + b_1 \frac{n_r}{n^2} erorr_l + b_2 \frac{n_c}{n} erorr_l \quad \text{(A4)}$$

where b_1, b_2 are the learning parameters, n_c is the number of the concepts in the candidate FCM model, n_r is the number of the non-zero relationships between concepts, n is the number of all possible concepts, and $erorr_l$ is the learning error type (Equation (A3)).

In this paper, we also used the hybrid approach for time series prediction based on fuzzy cognitive maps with the structure optimization genetic algorithm and artificial neural networks [101]. In the first stage of this approach, the most important concepts are selected with the use of FCMs and the SOGA algorithm. In the second stage, these concepts are used as inputs for the artificial neural network in order to increase the prediction accuracy. The above algorithms have been implemented in the developed ISEMK system [101].

References

1. Lee, Y.-S.; Tong, L.-I. Forecasting energy consumption using a grey model improved by incorporating genetic programming. *Energy Convers. Manag.* **2011**, *52*, 147–152. [CrossRef]
2. Barak, S.; Sadegh, S.S. Forecasting energy consumption using ensemble ARIMA–ANFIS hybrid algorithm. *Int. J. Electr. Power Energy Syst.* **2016**, *82*, 92–104. [CrossRef]
3. Adedeji, P.; Akinlabi, S.; Madushele, N.; Olatunji, O. Hybrid Adaptive Neuro-fuzzy Inference System (ANFIS) for a Multi-campus University Energy Consumption Forecast. *Int. J. Ambient Energy* **2020**, *41*, 1–20. [CrossRef]
4. Yu, S.; Wei, Y.-M.; Wang, K. A PSO-GA optimal model to estimate primary energy demand of China. *Energy Policy* **2012**, *42*, 329–340. [CrossRef]
5. U.S. Energy Information Administraton. International Energy Outlook. 2016. Available online: https://www.eia.gov/outlooks/ieo/pdf/0484(2016)pdf (accessed on 3 January 2020).
6. Wei, N.; Li, C.; Li, C.; Xie, H.; Du, Z.; Zhang, Q.; Zeng, F. Short-Term Forecasting of Natural Gas Consumption Using Factor Selection Algorithm and Optimized Support Vector Regression. *J. Energy Res. Technol.* **2018**, *141*, 032701. [CrossRef]
7. Tamba, J.G.; Ndjakomo, S.; Sapnken, E.; Koffi, F.; Nsouandele, J.L.; Soldo, B.; Njomo, D. Forecasting Natural Gas: A Literature Survey. *Int. J. Energy Econ. Policy* **2018**, *8*, 216–249.
8. Salehnia, N.; Falahi, M.A.; Seifi, A.; Adeli, M.H.M. Forecasting natural gas spot prices with nonlinear modeling using Gamma test analysis. *J. Nat. Gas Sci. Eng.* **2013**, *14*, 238–249. [CrossRef]
9. Xu, G.; Wang, W. Forecasting China's natural gas consumption based on a combination model. *J. Nat. Gas Chem.* **2010**, *19*, 493–496. [CrossRef]
10. Agbonifo, P.E. Natural Gas Distribution Infrastructure and the Quest for Environmental Sustainability in the Niger Delta: The Prospect of Natural Gas Utilization in Nigeria. *Int. J. Energy Econ. Policy* **2016**, *6*, 442–448.
11. Solarin, S.A.; Ozturk, I. The relationship between natural gas consumption and economic growth in OPEC members. *Renew. Sustain. Energy Rev.* **2016**, *58*, 1348–1356. [CrossRef]
12. Panapakidis, I.P.; Dagoumas, A.S. Day-ahead natural gas demand forecasting based on the combination of wavelet transform and ANFIS/genetic algorithm/neural network model. *Energy* **2017**, *118*, 231–245. [CrossRef]

13. Papageorgiou, K.; Papageorgiou, E.; Poczeta, K.; Gerogiannis, V.; Stamoulis, G. Exploring an Ensemble of Methods that Combines Fuzzy Cognitive Maps and Neural Networks in Solving the Time Series Prediction Problem of Gas Consumption in Greece. *Algorithms* **2019**, *12*, 235. [CrossRef]
14. Ediger, V.Ş.; Akar, S. ARIMA forecasting of primary energy demand by fuel in Turkey. *Energy Policy* **2007**, *35*, 1701–1708. [CrossRef]
15. Šebalj, D.; Mesaric, J.; Dujak, D. Predicting Natural Gas Consumption—A Literature Review. In Proceedings of the Central European Conference on Information and Intelligent Systems, Varaždin, Croatia, 27–29 September 2017.
16. Gil, S.; Deferrari, J. Generalized Model of Prediction of Natural Gas Consumption. *J. Energy Res. Technol.* **2004**, *126*, 90–98. [CrossRef]
17. Akpınar, M.; Yumuşak, N. Estimating household natural gas consumption with multiple regression: Effect of cycle. In Proceedings of the 2013 International Conference on Electronics, Computer and Computation, ICECCO 2013, Ankara, Turkey, 7–9 November 2013; pp. 188–191.
18. Akpinar, M.; Yumusak, N. Day-ahead natural gas forecasting using nonseasonal exponential smoothing methods. In Proceedings of the 2017 IEEE International Conference on Environment and Electrical Engineering and 2017 IEEE Industrial and Commercial Power Systems Europe (EEEIC/I CPS Europe), Milan, Italy, 6–9 June 2017; pp. 1–4.
19. Akpınar, M.; Yumuşak, N. Forecasting household natural gas consumption with ARIMA model: A case study of removing cycle. In Proceedings of the 2013 7th International Conference on Application of Information and Communication Technologies, Baku, Azerbaijan, 23–25 October 2013; pp. 1–6.
20. Deka, A.; Hamta, N.; Esmaeilian, B.; Behdad, S. Predictive Modeling Techniques to Forecast Energy Demand in the United States: A Focus on Economic and Demographic Factors. In Proceedings of the Journal of Energy Resources Technology, Boston, MA, USA, 2–5 August 2015; Volume 138.
21. Kaynar, O.; Yilmaz, I.; Demirkoparan, F. Forecasting of natural gas consumption with neural network and neuro fuzzy system. *Energy Educ. Sci. Technol. Part A Energy Sci. Res.* **2010**, *26*, 221–238.
22. Zhu, L.; Li, M.S.; Wu, Q.H.; Jiang, L. Short-term natural gas demand prediction based on support vector regression with false neighbours filtered. *Energy* **2015**, *80*, 428–436. [CrossRef]
23. Yu, F.; Xu, X. A short-term load forecasting model of natural gas based on optimized genetic algorithm and improved BP neural network. *Appl. Energy* **2014**, *134*, 102–113. [CrossRef]
24. Potočnik, P.; Govekar, E.; Grabec, I. Short-term natural gas consumption forecasting. In Proceedings of the IASTED International Conference on Applied Simulation and Modelling, ASM 2011, Palma de Mallorca, Spain, 29–31 August 2007; pp. 353–357.
25. Merkel, G.D.; Povinelli, R.J.; Brown, R.H. Short-Term Load Forecasting of Natural Gas with Deep Neural Network Regression. *Energies* **2018**, *11*, 2008. [CrossRef]
26. Pedregal, D.J.; Trapero, J.R. Mid-term hourly electricity forecasting based on a multi-rate approach. *Energy Convers. Manag.* **2010**, *51*, 105–111. [CrossRef]
27. Almeshaiei, E.; Soltan, H. A methodology for Electric Power Load Forecasting. *Alex. Eng. J.* **2011**, *50*, 137–144. [CrossRef]
28. Park, D.C.; El-Sharkawi, M.A.; Marks, R.J.; Atlas, L.E.; Damborg, M.J. Electric load forecasting using an artificial neural network. *IEEE Trans. Power Syst.* **1991**, *6*, 442–449. [CrossRef]
29. Pai, P.-F.; Hong, W.-C. Forecasting regional electricity load based on recurrent support vector machines with genetic algorithms. *Electr. Power Syst. Res.* **2005**, *74*, 417–425. [CrossRef]
30. Pai, P.-F.; Hong, W.-C. Support vector machines with simulated annealing algorithms in electricity load forecasting. *Energy Convers. Manag.* **2005**, *46*, 2669–2688. [CrossRef]
31. Hippert, H.S.; Pedreira, C.E.; Souza, R.C. Neural networks for short-term load forecasting: A review and evaluation. *IEEE Trans. Power Syst.* **2001**, *16*, 44–55. [CrossRef]
32. Hong, W.-C. Electric load forecasting by support vector model. *Appl. Math. Model.* **2009**, *33*, 2444–2454. [CrossRef]
33. Nizami, S.J.; Al-Garni, A.Z. Forecasting electric energy consumption using neural networks. *Energy Policy* **1995**, *23*, 1097–1104. [CrossRef]
34. Kalogirou, S.A. Applications of artificial neural networks in energy systems. *Energy Convers. Manag.* **1999**, *40*, 1073–1087. [CrossRef]

35. Aydinalp, M.; Ismet Ugursal, V.; Fung, A.S. Modeling of the appliance, lighting, and space-cooling energy consumptions in the residential sector using neural networks. *Appl. Energy* **2002**, *71*, 87–110. [CrossRef]
36. Hsu, C.-C.; Chen, C.-Y. Regional load forecasting in Taiwan—Applications of artificial neural networks. *Energy Convers. Manag.* **2003**, *44*, 1941–1949. [CrossRef]
37. Roldán-Blay, C.; Escrivá-Escrivá, G.; Álvarez-Bel, C.; Roldán-Porta, C.; Rodríguez-García, J. Upgrade of an artificial neural network prediction method for electrical consumption forecasting using an hourly temperature curve model. *Energy Build.* **2013**, *60*, 38–46. [CrossRef]
38. Raza, M.Q.; Baharudin, Z. A review on short term load forecasting using hybrid neural network techniques. In Proceedings of the 2012 IEEE International Conference on Power and Energy (PECon), Kota Kinabalu, Malaysia, 2–5 December 2012; pp. 846–851.
39. De Felice, M.; Yao, X. Notes Short-Term Load Forecasting with Neural Network Ensembles: A Comparative Study. *Comput. Intell. Mag.* **2011**, *6*, 47–56. [CrossRef]
40. Kumar, R.; Aggarwal, R.K.; Sharma, J.D. Energy analysis of a building using artificial neural network: A review. *Energy Build.* **2013**, *65*, 352–358. [CrossRef]
41. Sulaiman, S.; Jeyanthy, A.; Devaraj, D. Artificial neural network based day ahead load forecasting using Smart Meter data. In Proceedings of the 2016 Biennial International Conference on Power and Energy Systems: Towards Sustainable Energy (PESTSE), Bangalore, India, 21–23 January 2016; pp. 1–6.
42. Abdel-Aal, R.E.; Al-Garni, A.Z.; Al-Nassar, Y.N. Modelling and forecasting monthly electric energy consumption in eastern Saudi Arabia using abductive networks. *Energy* **1997**, *22*, 911–921. [CrossRef]
43. Kermanshahi, B. Recurrent neural network for forecasting next 10 years loads of nine Japanese utilities. *Neurocomputing* **1998**, *23*, 125–133. [CrossRef]
44. Ghelardoni, L.; Ghio, A.; Anguita, D. Energy Load Forecasting Using Empirical Mode Decomposition and Support Vector Regression. *IEEE Trans. Smart Grid* **2013**, *4*, 549–556. [CrossRef]
45. Ceylan, H.; Ozturk, H. Estimating energy demand of Turkey based on economic indicators using genetic algorithm approach. *Energy Convers. Manag.* **2004**, *45*, 2525–2537. [CrossRef]
46. Ozturk, H.K.; Ceylan, H.; Canyurt, O.E.; Hepbasli, A. Electricity estimation using genetic algorithm approach: A case study of Turkey. *Energy* **2005**, *30*, 1003–1012. [CrossRef]
47. Azadeh, A.; Ghaderi, S.F.; Tarverdian, S.; Saberi, M. Integration of artificial neural networks and genetic algorithm to predict electrical energy consumption. *Appl. Math. Comput.* **2007**, *186*, 1731–1741. [CrossRef]
48. Daut, M.A.M.; Hassan, M.Y.; Abdullah, H.; Rahman, H.A.; Abdullah, M.P.; Hussin, F. Building electrical energy consumption forecasting analysis using conventional and artificial intelligence methods: A review. *Renew. Sustain. Energy Rev.* **2017**, *70*, 1108–1118. [CrossRef]
49. Ahmad, T.; Chen, H.; Guo, Y.; Wang, J. A comprehensive overview on the data driven and large scale based approaches for forecasting of building energy demand: A review. *Energy Build.* **2018**, *165*, 301–320. [CrossRef]
50. Foucquier, A.; Robert, S.; Suard, F.; Stéphan, L.; Jay, A. State of the art in building modelling and energy performances prediction: A review. *Renew. Sustain. Energy Rev.* **2013**, *23*, 272–288. [CrossRef]
51. Metaxiotis, K.; Kagiannas, A.; Askounis, D.; Psarras, J. Artificial intelligence in short term electric load forecasting: A state-of-the-art survey for the researcher. *Energy Convers. Manag.* **2003**, *44*, 1525–1534. [CrossRef]
52. GORUCU, F.B. Artificial Neural Network Modeling for Forecasting Gas Consumption. *Energy Sources* **2004**, *26*, 299–307. [CrossRef]
53. Karimi, H.; Dastranj, J. Artificial neural network-based genetic algorithm to predict natural gas consumption. *Energy Syst.* **2014**, *5*, 571–581. [CrossRef]
54. Khotanzad, A.; Elragal, H.M. Natural gas load forecasting with combination of adaptive neural networks. In Proceedings of the IJCNN'99. International Joint Conference on Neural Networks. Proceedings (Cat. No. 99CH36339), Washington, DC, USA, 10–16 July 1999; Volume 6, pp. 4069–4072.
55. Khotanzad, A.; Elragal, H.; Lu, T. Combination of artificial neural-network forecasters for prediction of natural gas consumption. *IEEE Trans. Neural Netw.* **2000**, *11*, 464–473. [CrossRef]
56. Kizilaslan, R.; Karlik, B. Combination of neural networks forecasters for monthly natural gas consumption prediction. *Neural Netw. World* **2009**, *19*, 191–199.

57. Kizilaslan, R.; Karlik, B. Comparison neural networks models for short term forecasting of natural gas consumption in Istanbul. In Proceedings of the 2008 First International Conference on the Applications of Digital Information and Web Technologies (ICADIWT), Ostrava, Czech Republic, 4–6 August 2008, pp. 448–453.
58. Musílek, P.; Pelikán, E.; Brabec, T.; Simunek, M. Recurrent Neural Network Based Gating for Natural Gas Load Prediction System. In Proceedings of the 2006 IEEE International Joint Conference on Neural Network Proceedings, Vancouver, BC, Canada, 16–21 July 2006; pp. 3736–3741.
59. Soldo, B. Forecasting natural gas consumption. *Appl. Energy* **2012**, *92*, 26–37. [CrossRef]
60. Szoplik, J. Forecasting of natural gas consumption with artificial neural networks. *Energy* **2015**, *85*, 208–220. [CrossRef]
61. Aydinalp-Koksal, M.; Ugursal, V.I. Comparison of neural network, conditional demand analysis, and engineering approaches for modeling end-use energy consumption in the residential sector. *Appl. Energy* **2008**, *85*, 271–296. [CrossRef]
62. Dombayci, Ö. The prediction of heating energy consumption in a model house by using artificial neural networks in Denizli-Turkey. *Adv. Eng. Softw.* **2010**, *41*, 141–147. [CrossRef]
63. Taşpınar, F.; Çelebi, N.; Tutkun, N. Forecasting of daily natural gas consumption on regional basis in Turkey using various computational methods. *Energy Build.* **2013**, *56*, 23–31. [CrossRef]
64. Lago, J.; Ridder, F.D.; Schutter, B.D. Forecasting spot electricity prices: Deep learning approaches and empirical comparison of traditional algorithms. *Appl. Energy* **2018**, *221*, 386–405. [CrossRef]
65. Tonkovic, Z.; Zekic-Susac, M.; Somolanji, M. Predicting natural gas consumption by neural networks. *Teh. Vjesn.* **2009**, *16*, 51–61.
66. Demirel, F.; Zaim, S.; Caliskan, A.; Gokcin Ozuyar, P. Forecasting natural gas consumption in Istanbul using neural networks and multivariate time series methods. *Turk. J. Electr. Eng. Comput. Sci.* **2012**, *20*, 695–711. [CrossRef]
67. Olgun, M.; Ozdemir, G.; Aydemir, E. Forecasting of Turkey's natural gas demand using artifical neural networks and support vector machines. *Energy Educ. Sci. Technol.* **2011**, *30*, 15–20.
68. Soldo, B.; Potočnik, P.; Šimunović, G.; Šarić, T.; Govekar, E. Improving the residential natural gas consumption forecasting models by using solar radiation. *Energy Build.* **2014**, *69*, 498–506. [CrossRef]
69. Izadyar, N.; Ong, H.C.; Shamshirband, S.; Ghadamian, H.; Tong, C.W. Intelligent forecasting of residential heating demand for the District Heating System based on the monthly overall natural gas consumption. *Energy Build.* **2015**, *104*, 208–214. [CrossRef]
70. Ivezić, D. Short-term natural gas consumption forecast. *FME Trans.* **2006**, *34*, 165–169.
71. Garcia, A.; Mohaghegh, S.D. Forecasting US Natural Gas Production into year 2020: A comparative study. In Proceedings of the SPE Eastern Regional Meeting, Charleston, WV, USA, 15–17 September 2004; Society of Petroleum Engineers: Richardson, TX, USA, 2004; p. 5.
72. Ma, H.; Wu, Y. Grey Predictive on Natural Gas Consumption and Production in China. In Proceedings of the 2009 Second Pacific-Asia Conference on Web Mining and Web-based Application, Wuhan, China, 6–7 June 2009; pp. 91–94.
73. Ervural, B.C.; Beyca, O.F.; Zaim, S. Model Estimation of ARMA Using Genetic Algorithms: A Case Study of Forecasting Natural Gas Consumption. *Procedia Soc. Behav. Sci.* **2016**, *235*, 537–545. [CrossRef]
74. Zeng, B.; Li, C. Forecasting the natural gas demand in China using a self-adapting intelligent grey model. *Energy* **2016**, *112*, 810–825. [CrossRef]
75. Liu, G.; Dong, X.; Jiang, Q.; Dong, C.; Li, J. Natural gas consumption of urban households in China and corresponding influencing factors. *Energy Policy* **2018**, *122*, 17–26. [CrossRef]
76. Wang, D.; Liu, Y.; Wu, Z.; Fu, H.; Shi, Y.; Guo, H. Scenario Analysis of Natural Gas Consumption in China Based on Wavelet Neural Network Optimized by Particle Swarm Optimization Algorithm. *Energies* **2018**, *11*, 825. [CrossRef]
77. Hribar, R.; Potočnik, P.; Šilc, J.; Papa, G. A comparison of models for forecasting the residential natural gas demand of an urban area. *Energy* **2019**, *167*, 511–522. [CrossRef]
78. Laib, O.; Khadir, M.T.; Mihaylova, L. Toward efficient energy systems based on natural gas consumption prediction with LSTM Recurrent Neural Networks. *Energy* **2019**, *177*, 530–542. [CrossRef]
79. Chen, Y.; Chua, W.S.; Koch, T. Forecasting day-ahead high-resolution natural-gas demand and supply in Germany. *Appl. Energy* **2018**, *228*, 1091–1110. [CrossRef]

80. Beyca, O.F.; Ervural, B.C.; Tatoglu, E.; Ozuyar, P.G.; Zaim, S. Using machine learning tools for forecasting natural gas consumption in the province of Istanbul. *Energy Econ.* **2019**, *80*, 937–949. [CrossRef]
81. Ding, S. A novel self-adapting intelligent grey model for forecasting China's natural-gas demand. *Energy* **2018**, *162*, 393–407. [CrossRef]
82. Fan, G.-F.; Wang, A.; Hong, W.-C. Combining Grey Model and Self-Adapting Intelligent Grey Model with Genetic Algorithm and Annual Share Changes in Natural Gas Demand Forecasting. *Energies* **2018**, *11*, 1625. [CrossRef]
83. Brown, R.G.; Matin, L.; Kharout, P.; Piessens, L.P. Development of artificial neural-network models to predict daily gas consumption. In Proceedings of the IECON '95—21st Annual Conference on IEEE Industrial Electronics, Orlando, FL, USA, 6–10 November 1996; Volume 5, pp. 1–22.
84. Viet, N.H.; Mandziuk, J. Neural and fuzzy neural networks for natural gas consumption prediction. In Proceedings of the 2003 IEEE XIII Workshop on Neural Networks for Signal Processing (IEEE Cat. No.03TH8718), Toulouse, France, 17–19 September 2003; pp. 759–768.
85. Ying, L.-C.; Pan, M.-C. Using adaptive network based fuzzy inference system to forecast regional electricity loads. *Energy Convers. Manag.* **2008**, *49*, 205–211. [CrossRef]
86. Akdemir, B.; Çetinkaya, N. Long-term load forecasting based on adaptive neural fuzzy inference system using real energy data. *Energy Procedia* **2012**, *14*, 794–799. [CrossRef]
87. Mordjaoui, M.; Boudjema, B. Forecasting and Modelling Electricity Demand Using Anfis Predictor. *J. Math. Stat.* **2011**, *7*, 275–281. [CrossRef]
88. Azadeh, A.; Saberi, M.; Gitiforouz, A.; Saberi, Z. A hybrid simulation adaptive-network-based fuzzy inference system for improvement of electricity consumption estimation. *Expert Syst. Appl.* **2009**, *36*, 11108–11117. [CrossRef]
89. Cheng, C.-H.; Wei, L.-Y. One step-ahead ANFIS time series model for forecasting electricity loads. *Optim. Eng.* **2010**, *11*, 303–317. [CrossRef]
90. Senvar, O.; Ulutagay, G.; Turanoğlu Bekar, E. ANFIS Modeling for Forecasting Oil Consumption of Turkey. *J. Mult.-Valued Log. Soft Comput.* **2016**, *26*, 609–624.
91. Azadeh, A.; Asadzadeh, S.M.; Ghanbari, A. An adaptive network-based fuzzy inference system for short-term natural gas demand estimation: Uncertain and complex environments. *Energy Policy* **2010**, *38*, 1529–1536. [CrossRef]
92. Zhang, Y.; Ling, C. A strategy to apply machine learning to small datasets in materials science. *Npj Comput. Mater.* **2018**, *4*, 25. [CrossRef]
93. Behrouznia, A.; Saberi, M.; Azadeh, A.; Asadzadeh, S.M.; Pazhoheshfar, P. An adaptive network based fuzzy inference system-fuzzy data envelopment analysis for gas consumption forecasting and analysis: The case of South America. In Proceedings of the 2010 International Conference on Intelligent and Advanced Systems, Manila, Philippines, 15–17 June 2010; pp. 1–6.
94. Azadeh, A.; Saberi, M.; Asadzadeh, S.M.; Hussain, O.K.; Saberi, Z. A neuro-fuzzy-multivariate algorithm for accurate gas consumption estimation in South America with noisy inputs. *Int. J. Electr. Power Energy Syst.* **2013**, *46*, 315–325. [CrossRef]
95. Azadeh, A.; Asadzadeh, S.M.; Saberi, M.; Nadimi, V.; Tajvidi, A.; Sheikalishahi, M. A Neuro-fuzzy-stochastic frontier analysis approach for long-term natural gas consumption forecasting and behavior analysis: The cases of Bahrain, Saudi Arabia, Syria, and UAE. *Appl. Energy* **2011**, *88*, 3850–3859. [CrossRef]
96. Azadeh, A.; Zarrin, M.; Rahdar Beik, H.; Aliheidari Bioki, T. A neuro-fuzzy algorithm for improved gas consumption forecasting with economic, environmental and IT/IS indicators. *J. Pet. Sci. Eng.* **2015**, *133*, 716–739. [CrossRef]
97. Salmeron, J.L.; Froelich, W. Dynamic optimization of fuzzy cognitive maps for time series forecasting. *Knowl.-Based Syst.* **2016**, *105*, 29–37. [CrossRef]
98. Froelich, W.; Salmeron, J.L. Evolutionary learning of fuzzy grey cognitive maps for the forecasting of multivariate, interval-valued time series. *Int. J. Approx. Reason.* **2014**, *55*, 1319–1335. [CrossRef]
99. Papageorgiou, E.I.; Poczeta, K.; Laspidou, C. Application of Fuzzy Cognitive Maps to water demand prediction. In Proceedings of the IEEE International Conference on Fuzzy Systems, Istanbul, Turkey, 2–5 August 2015; Volume 2015.

100. Poczeta, K.; Yastrebov, A.; Papageorgiou, E.I. Learning fuzzy cognitive maps using structure optimization genetic algorithm. In Proceedings of the 2015 Federated Conference on Computer Science and Information Systems, FedCSIS 2015, Lodz, Poland, 13–16 September 2015.
101. Papageorgiou, E.I.; Poczeta, K. A two-stage model for time series prediction based on fuzzy cognitive maps and neural networks. *Neurocomputing* **2017**, *232*, 113–121. [CrossRef]
102. Poczeta, K.; Papageorgiou, E.I. Implementing Fuzzy Cognitive Maps with Neural Networks for Natural Gas Prediction. In Proceedings of the 2018 IEEE 30th International Conference on Tools with Artificial Intelligence (ICTAI), Volos, Greece, 5–7 November 2018; pp. 1026–1032.
103. Hellenic Gas Transmission System Operator S.A. (DESFA). Available online: https://www.desfa.gr/en/ (accessed on 1 March 2020).
104. Dudek, G. Multilayer perceptron for short-term load forecasting: From global to local approach. *Neural Comput. Appl.* **2020**, *32*, 3695–3707. [CrossRef]
105. Azadeh, A.; Ghaderi, S.F.; Sohrabkhani, S. A simulated-based neural network algorithm for forecasting electrical energy consumption in Iran. *Energy Policy* **2008**, *36*, 2637–2644. [CrossRef]
106. Takagi, T.; Sugeno, M. Fuzzy identification of systems and its applications to modeling and control. *IEEE Trans. Syst. Man Cybern.* **1985**, *SMC-15*, 116–132. [CrossRef]
107. Jang, J.-R. ANFIS: Adaptive-network-based fuzzy inference system. *IEEE Trans. Syst. Man Cybern.* **1993**, *23*, 665–685. [CrossRef]
108. Rosadi, D.; Subanar, T. Suhartono Analysis of Financial Time Series Data Using Adaptive Neuro Fuzzy Inference System (ANFIS). *Int. J. Comput. Sci. Issues* **2013**, *10*, 491–496.
109. Jang, J.-S.; Sun, C.-T.; Mizutani, C.T. Neuro-Fuzzy and Soft Computing—A Computational Approach to Learning and Machine Intelligence. In *Prentice Hall Upper Saddle River*; IEEE: Piscataway, NJ, USA, 1997; Volume 42, pp. 1482–1484.
110. Jang, J.-S.; Sun, C.-T. Neuro-Fuzzy Modeling and Control. *Proc. IEEE* **1995**, *83*, 378–406. [CrossRef]
111. Yeom, C.-U.; Kwak, K.-C. Performance Comparison of ANFIS Models by Input Space Partitioning Methods. *Symmetry* **2018**, *10*, 700. [CrossRef]
112. Naderloo, L.; Alimardani, R.; Omid, M.; Sarmadian, F.; Javadikia, P.; Torabi, M.Y.; Alimardani, F. Application of ANFIS to predict crop yield based on different energy inputs. *Measurement* **2012**, *45*, 1406–1413. [CrossRef]
113. Papageorgiou, E.I.; Aggelopoulou, K.; Gemtos, T.A.; Nanos, G.D. Development and Evaluation of a Fuzzy Inference System and a Neuro-Fuzzy Inference System for Grading Apple Quality. *Appl. Artif. Intell.* **2018**, *32*, 253–280. [CrossRef]
114. Papageorgiou, E.I.; Groumpos, P.P. Two-stage learning algorithm for fuzzy cognitive maps. In Proceedings of the 2004 2nd International IEEE Conference "Intelligent Systems"—Proceedings, Varna, Bulgaria, 22–24 June 2004; Volume 1.
115. Papageorgiou, E.I.; Poczęta, K.; Laspidou, C. Hybrid model for water demand prediction based on fuzzy cognitive maps and artificial neural networks. In Proceedings of the 2016 IEEE International Conference on Fuzzy Systems, FUZZ-IEEE, Vancouver, BC, Canada, 24–29 July 2016.
116. Poczeta, K.; Kubuś, L.; Yastrebov, A.; Papageorgiou, E.I. *Temperature Forecasting for Energy Saving in Smart Buildings Based on Fuzzy Cognitive Map*; Springer: Berlin/Heidelberg, Germany, 2018; Volume 743, ISBN 978-3-319-77178-6.
117. Anagnostis, A.; Papageorgiou, E.; Dafopoulos, V.; Bochtis, D. Applying Long Short-Term Memory Networks for natural gas demand prediction. In Proceedings of the 2019 10th International Conference on Information, Intelligence, Systems and Applications (IISA), Patras, Greece, 15–17 July 2019; pp. 1–7.
118. Azizi, A. An adaptive neuro-fuzzy inference system for a dynamic production environment under uncertainties. *World Appl. Sci. J.* **2013**, *25*, 428–433. [CrossRef]
119. Raju, G.S.V.P.; Mary Sumalatha, V.; Ramani, K.V.; Lakshmi, K.V. Solving Uncertain Problems using ANFIS. *Int. J. Comput. Appl.* **2011**, *29*, 14–21. [CrossRef]
120. Poczeta, K.; Yastrebov, A.; Papageorgiou, E.I. *Forecasting Indoor Temperature Using Fuzzy Cognitive Maps with Structure Optimization Genetic Algorithm*; Springer: Berlin/Heidelberg, Germany, 2016; Volume 655.

energies

Article

Dry Above Ground Biomass for a Soybean Crop Using an Empirical Model in Greece

Christos Vamvakoulas, Stavros Alexandris and Ioannis Argyrokastritis *

Laboratory of Agricultural Hydraulics, Department of Natural Resources Development and Agricultural Engineering, Agricultural University of Athens, 75 Iera Odos str., 11855 Athens, Greece; chrisvamva@aua.gr (C.V.); stalex@aua.gr (S.A.)
* Correspondence: jarg@aua.gr

Received: 1 November 2019; Accepted: 24 December 2019; Published: 1 January 2020

Abstract: A new empirical equation for the estimation of daily dry above ground biomass (D-AGB) for a hybrid of soybean (*Glycine max* L.) is proposed. This equation requires data for three crop dependent parameters; leaf area index, plant height, and cumulative crop evapotranspiration. Bilinear surface regression analysis was used in order to estimate the factors entering in the empirical model. For the calibration of the proposed model, data yielded from a well-watered soybean crop for the year 2015, in the experimental field (0.1 ha) of the agricultural University of Athens, were used as a reference. Verification of the validity of the model was obtained by using data from a 2014 cultivation period for well-watered soybean cultivation (100% of crop evapotranspiration water treatment), as well as data from three irrigation treatments (75%, 50%, 25% of crop evapotranspiration) for two cultivation periods (2014–2015). The proposed method for the estimation of D-AGB may be proven as a useful tool for estimations without using destructive sampling.

Keywords: dry above ground biomass; soybean; empirical models; bilinear regression analysis

---

1. Introduction

Nowadays, agronomists and irrigation experts use crop productivity models for the simulation and prediction of dry above ground biomass (D-AGB). Complex crop growth models (CGM) require a large number of input parameters, usually not available from ideal sites, which leads to significant and systematic cumulative errors in determining crop yield and above ground biomass.

In this study we tried to present a simple model using geostatistical methods in a simple form. A similar approach is the responsive surface methodology (RSM), which has been used by researchers as an optimization technique with a wide range of applications, such as in dehydration operations of selected agricultural crops [1] and for biodiesel production [2–6]. Also, in the optimization of tomato yield [7], to increase resistant starch in vermicelli [8] and optimization of the plough working surface [9]. Other fields of use for this optimization method are in the use of agricultural waste [10] and for harvesting losses in corn seed [11].

Generally, empirical equations are a tool for local estimations of attributes without many parameters as inputs. In the past, algorithms for the creation of such equations have been used for the estimation of reference evapotranspiration (ET_0) [12,13] and for the estimation of crop evapotranspiration [14].

Also, empirical bilinear regression equations have been used for the prediction of human and rat tissue [15], while multiple regression analysis has been applied for un-mixing of surface temperature data in an urban environment [16].

In this paper, a model for the daily estimation of D-AGB for a soybean hybrid (PR91M10) in central Greece is formulated. The model has been parameterized by experimental observations on the soybean crop. Also, the model was examined for water stressed and non-water stressed plants, under field conditions. The final equation obtained is based on leaf area index (LAI), plant height (h_c), and cumulative crop evapotranspiration (cumET$_c$).

2. Methodology

The experiment was performed in the experimental field of the agricultural University of Athens in Aliartos plain (38°23′40″ N, 23°05′08″ E and 95 m altitude), during 2014 and 2015 cultivation periods. Data from an experimental design with irrigation treatments (100%, 75%, 50%, 25% of crop evapotranspiration, respectively) as the main plot in a randomized complete block design, with four replications, were used. The plot size of each irrigation treatment was 3 m × 12 m and the spacing between each main plot was 3 m in order to minimize water movement among treatments. The experimental plots were 3 m × 6 m and consisted of 5 rows 0.75 m apart. PR91M10 is a highly productive variety of the early maturity group (00). Seeds were hand-planted, using a seeding depth of about 3 cm, on 30 May 2014 (Julian day, JD: 150) and on 31 May 2015 (Julian day, JD: 151), respectively. Treatment plots consisted of 5 rows planted, 75 cm apart, with 4–5 cm row spacing, and the sowing density was 33 seeds m^{-2}. At sowing, basal fertilization was performed in all plots and corresponded to 100 and 50 kg ha^{-1} of P_2O_5 and K_2O, respectively, based on soil fertility analysis. Irrigation scheduling was based on the daily water balance calculation and on results obtained by using the computer model ISAREG [17], which utilized data collected during consecutive cultivation periods from 2011 to 2015. The following input data for the ISAREG model were used:

(1) Daily grass reference evapotranspiration as estimated by the FAO56 Penman-Monteith equation [18]. All the meteorological parameters used were collected from the automatic meteorological station of the laboratory of agricultural hydraulics, which is installed on a well-watered extended grass field, very close to the experimental plots (100 m). Meteorological data, such as air temperature (T_{avg}, T_{max}, and T_{min}), air relative humidity (RH_{avg}, RH_{mean}, RH_{max}), wind speed at the level 2 m (u_2), solar radiation (Rs), net radiation (R_{net}), photosynthetically active radiation sensor (PAR), and soil temperature (T_{soil}), were collected. A rain gauge and wind direction sensor were also installed. All data were automatically collected and recorded from an acquisition system (data logger Campbell CR10X) in hourly and daily time step (averages).

(2) Soil data as determined from the 1 m soil profile of the experimental field, which was characterized as clay (sand 21%, clay 60%, silt 19%), with a field capacity of 0.43 m^3·m^{-3} and a wilting point of 0.15 m^3·m^{-3}.

(3) The adjusted K_c values, which were $K_{c,ini}$: 0.47, $K_{c,mid}$: 1.10, and $K_{c,end}$: 0.50. The adjusted K_c values were estimated by using the single crop coefficient K_c method [18].

Rainfall during the 2014 cultivation period was 46.1 mm and 176.7 mm for the 2015 cultivation period. The ground water table was at 1.2 m depth for both cultivation years. Irrigation was applied to provide 100%, 75%, 50%, and 25% of the crop evapotranspiration needs.

A surface drip irrigation system was used for irrigation. A 16 mm diameter polyethylene pipe with inline pressure compensating drippers at 33 cm intervals was placed on one side of each soybean row. The average discharge of emitters was 4.4 L/h at 0.1 MPa.

Periodically, every 7 days approximately, plant height (h_c, cm) was measured and destructive sampling was performed by collecting three plants from the three interior rows of each plot, for leaf area index (LAI) and dry above ground biomass (D-AGB, ton/ha). Sampling was performed at 25, 34, 41, 48, 54, 60, 66, and 75 days after planting (DAP) for 2014 cultivation period and at 24, 33, 40, 47, 53, 59, 65, and 75 (DAP) for the 2015 cultivation period, respectively.

The parameterization of the model was done for the 2015 cultivation year data, because precipitation was higher than that of 2014, giving better environmental conditions for the non-water stressed plants (100% treatment). The model represents the simulation curve of the D-AGB for the first 75 days of the growing period. The last 20 days of the maturity stage are not included in the simulation curve. Surface regression analysis was used to establish the new model to simulate daily (D-AGB). The empirical model was derived by surface polynomial regression using three crop dependent parameters, measured values of leaf area index (LAI), plant height (h_c), and cumulative crop evapotranspiration (cumET$_c$), in a general form D-AGB = f(LAI,h_c, cumET$_c$). It utilizes four unknown parameters (k_0, k_1, k_2, k_3) which are determined in a three stage approach. Experimental lines for the D-AGB obtained from the destructive sampling performed are used as standard values. Calculated D-AGB values are then regressed against mean daily values of pairs of LAI and h_c (first stage) and LAI and cumET$_c$ (second stage) in a bilinear equation of the form:

$$z = f(x, y) = k_0 + k_1 \cdot y + k_2 \cdot x + k_3 \cdot x \cdot y$$

x, y denoting daily values of either LAI and h_c (cm), in the first stage of investigation, or LAI and cumET$_c$ (mm/time), in the second stage, z standing for D-AGB (ton/ha). As expected, the first and second stages end up with the estimation of two sets of four parameters a_i, b_i ($i = 1, \ldots 4$), as shown in the Equations (1) and (2) below:

$$C_1 = a_1 + a_2 \cdot h_c + a_3 \cdot \text{LAI} + a_4 \cdot \text{LAI} \cdot h_c \quad (1)$$

where $a_1 = -0.143$, $a_2 = 0.095$, $a_3 = -6.33$, $a_4 = 0.058$.

$$C_2 = b_1 + b_2 \cdot \text{cumET}_c + b_3 \cdot \text{LAI} + b_4 \cdot \text{LAI} \cdot \text{cumET}_c \quad (2)$$

where $b_1 = -0.115$, $b_2 = 0.0066$, $b_3 = -2.4$, $b_4 = 0.0129$.

In the above equations C_1 and C_2 represent dry above ground biomass (D-AGB) in ton/ha. Tables 1 and 2 show the cross-correlation/covariance of the factors entering in the first and second stage of regression, respectively.

The D-AGB values are now regressed against the results obtained from the previous stages shown as C_1 and C_2 bilinear expressions (stage 3). This last regression ends up with the estimation of four parameters $m_i (i = 1, \ldots 4)$ and the final working formula for D-AGB on a daily basis is given by the following Equation (3) in an implicit form, since C_1 and C_2 are functions of the attributes LAI, h_c, and cumET$_c$:

$$\text{D-AGB(ton/ha)} = m_1 + m_2 \cdot C_2 + m_3 \cdot C_1 + m_4 \cdot C_1 \cdot C_2 \quad (3)$$

where $m_1 = 0.0082$, $m_2 = 1.11$, $m_3 = -0.12$, $m_4 = 0.0032$.

The above algorithm is exemplified by a flowchart diagram shown in Figure 1.

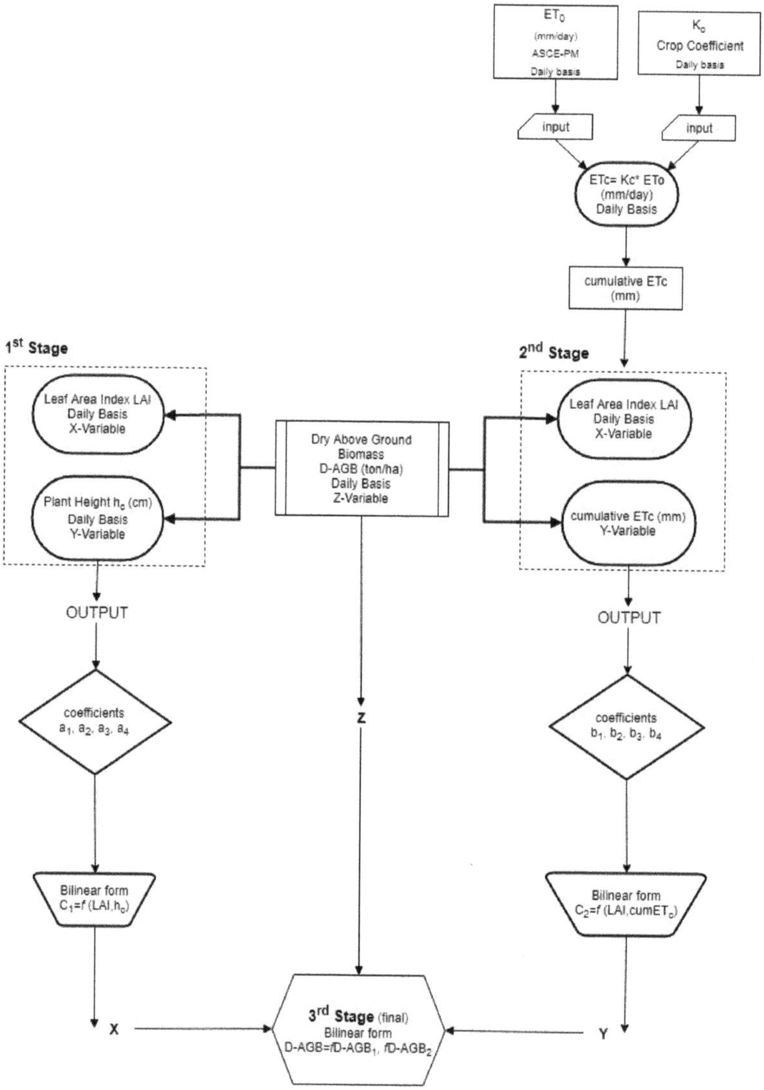

Figure 1. Flow chart showing the procedure of establishing the new empirical equation for daily estimation of dry above ground biomass (D-AGB).

In Figure 2a, the iso-lines of (D-AGB) derived from Equation (1) as a function of LAI, plant height (h_c), and D-AGB measurements, through curve interpolation lines, respectively, on a daily basis, are presented. Similarly, Figure 2b shows the results of the second stage D-AGB = f(LAI,cumET$_c$). Higher sensitivity showed the LAI-h$_c$ correlation rather than the one of LAI-cumET$_c$ for the D-AGB factor.

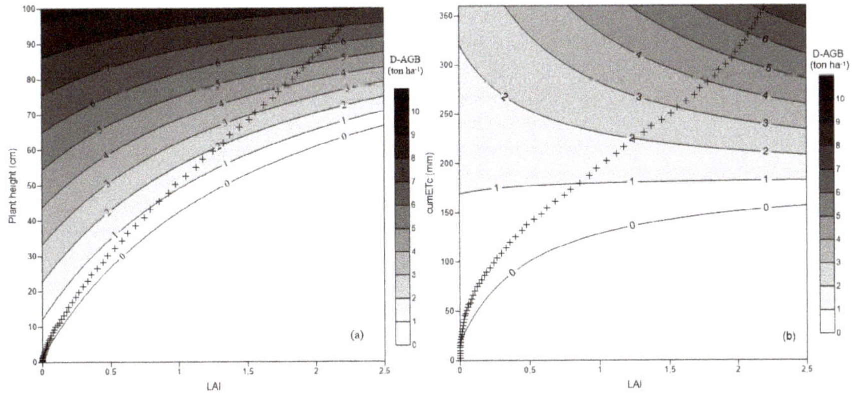

Figure 2. The iso-lines of (D-AGB) derived from Equations (1) and (2). (**a**) first stage (**b**) second stage.

Tables 1 and 2 show the cross-correlation/covariance of the factors entering in the first and second stages of regression, respectively.

Table 1. Cross-correlation/covariance between LAI, h_c, and D-AGB from the first stage of regression.

Cross-Correlation/Covariance	LAI	h_c	D-AGB
Variable correlation			
LAI	1.000	0.996	0.957
h_c	0.996	1.000	0.935
D-AGB	0.957	0.935	1.000
Variable covariance			
LAI	0.614	25.637	1.613
h_c	25.637	1080.251	66.108
D-AGB	1.613	66.108	4.626

Table 2. Cross-correlation/covariance between LAI, cumET$_c$, and D-AGB from the second stage of regression.

Cross-Correlation/Covariance	LAI	cumET$_c$	D-AGB
Variable correlation			
LAI	1.000	0.987	0.957
cumET$_c$	0.987	1.000	0.929
D-AGB	0.957	0.929	1.000
Variable covariance			
LAI	0.614	86.497	1.613
cumET$_c$	86.497	12,505.448	223.44
D-AGB	1.613	223.44	4.626

It is obvious from the Tables 1 and 2 that the strongest correlation exists between LAI and h_c (0.996) and that all three attributes; LAI, h_c, and cumET$_c$, are also strongly correlated to D-AGB (all correlation coefficients are above 0.92, see Tables 1 and 2).

The regression equations between daily simulated D-AGB values against the experimental and the cross-correlation coefficient (R^2) are shown in Table 3 for the 2015 and 2014 cultivation periods, respectively.

A statistical analysis was further performed in order to provide quantitative indices to our estimates. For this purpose, the following statistical indices were evaluated [19,20]: (i) Mean bias error (MBE), (ii) Variance of the distribution of differences s_d^2 which expresses the variability of $(P-O)$ distribution about MBE, (iii) Root mean square error (RMSE), (iv) Mean absolute error (MAE), (v) Index of agreement, d, [20], where n is the number of cases. O denotes the experimental values of D-AGB measured during the 2014–2015 cultivation periods for all irrigation treatments (I_{100}, I_{75}, I_{50}, and I_{25}). P denotes the simulated values as these are estimated by the proposed methodology. All the above mentioned relevant statistical indices are provided in Table 3.

Table 3. Summary statistics of daily dry above ground biomass (D-AGB) tested against the reference method.

Treatments	Slope	MBE	RMSE	MAE	s_d^2	d	R^2
2015, (N = 75)							
I_{75}	1.113	0.239	0.450	0.262	0.640	0.998	0.986
I_{50}	1.073	0.315	0.498	0.371	1.010	0.996	0.966
I_{25}	1.347	0.446	0.678	0.490	1.990	0.988	0.978
2014, (N = 75)							
I_{100}	1.213	0.226	0.380	0.245	0.539	0.997	0.992
I_{75}	1.211	0.393	0.579	0.414	1.522	0.994	0.974
I_{50}	1.008	0.211	0.378	0.321	0.485	0.998	0.965

3. Results and Discussion

In Figure 3, the development of D-AGB, both measured and simulated during cultivation period 2015 expressed in days after planting (DAP), is presented. As it is depicted in Figure 3a, the simulated and experimental curve interpolated lines almost coincided for the 100% treatment because the model has been calibrated for this treatment and cultivation period. Figure 3b–d depicts the 75%, 50%, and 25% treatments for the 2015 cultivation period, respectively. It is obvious that the predictions by the model for the 75%, 50%, and 25% treatments give results very close to the measurements for the 2015 cultivation period.

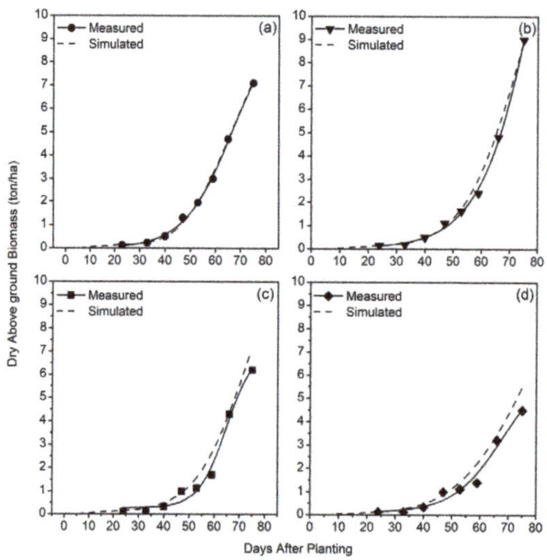

Figure 3. Relationship between dry above ground biomass (D-AGB), (ton/ha), and days after planting (D-AGB) for 2015 cultivation period and PR91M10 hybrid. The **(a)**, **(b)**, **(c)**, and **(d)** parts depict the 100%, 75%, 50%, and 25% of ET_c water treatments, respectively.

In Figure 4, the development of D-AGB both measured and simulated curve interpolated lines during cultivation period 2014 expressed in days after planting (DAP), are presented. For the 2014 cultivation period, all four figures (Figure 4a–d)were used for verification purposes. Figure 4a–d depicts the 100%, 75%, 50%, and 25% treatments, respectively. From Figure 4a at 100% treatment, it can be assumed that measured and simulated curve interpolated lines were very close, and at DAP 75 the model predicted D-AGB 4.951 ton/ha, while experimental D-AGB for the DAP 75 for the non water stressed soybean was 4.385 ton/ha. Similarly, the 75%, 50%, and 25% water treatments were perfect fitted until 55 DAP, approximately, for the 2014 cultivation period. However, the response of the plant to the water stress mechanism is a fairly complex process involving both biophysical and biochemical functions that could differentiate predictions of the experimental observations. This induces the differences after DAP 55 for the water stressed treatments in 2014 cultivation period (Figure 4b–d) and for the 2015 water stressed treatments (Figure 3b–d).

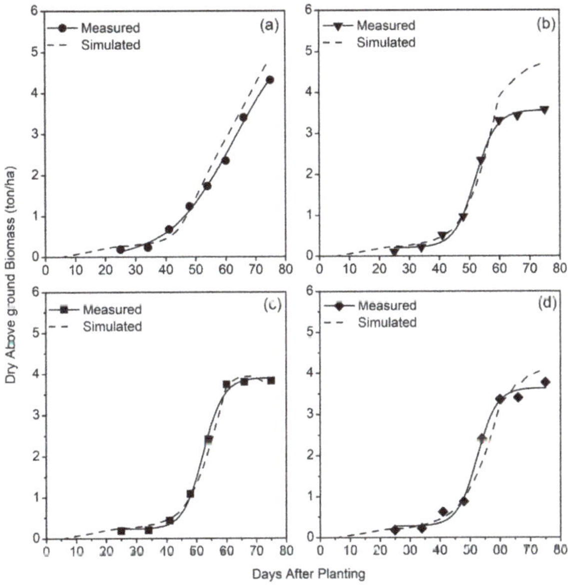

Figure 4. The relationship between dry above ground biomass (D-AGB), (ton/ha), and days after planting (DAP) for 2014 cultivation period and PR91M10 hybrid. The (**a**), (**b**), (**c**), and (**d**) parts depict the 100%, 75%, 50%, and 25% of ET_c water treatments, respectively.

4. Conclusions

For the first time, an already existing empirical methodology for the prediction of reference evapotranspiration (ET_0) used with crop geometrical characteristics (LAI, h_c) and cumET_c as inputs was used in order to predict daily D-AGB. The statistical analysis showed very satisfactory adjustment of the experimental and simulated values, especially for the non-water stressed treatments of the 2014 cultivation period.

Further experimentation for different regions and a wider range of D-AGB values is needed in order to verify the goodness of fit, for the parameters used in the methodology, in different climate regimes, and for more cultivation species. An important advantage of the methodology that has been followed, in addition to the use of three readily measured fundamental parameters (cumET_c, LAI, h_c), is that the model can easily be calibrated (different coefficients) for any crop and in any climatic environment.

However, a more complex algorithm could be set using more environmental attributes of the soil-plant-atmosphere system, which might be adjusted better to the simulated values than the experimental ones, especially for the plants under non-water stress or under irrigation deficit.

Author Contributions: Conceptualization (S.A. and I.A.); Investigation (C.V.); Methodology (S.A., I.A., C.V.); Supervision (I.A.); Writing—original draft (C.V.); Writing—review & editing (I.A., S.A., C.V.). All authors have read and agreed to the published version of the manuscript.

Funding: This research received no external funding.

Acknowledgments: This research was supported by the Alexander S. Onassis Foundation.

Conflicts of Interest: The authors declare no conflict of interest.

References

1. Madamba, S.P. The response surface methodology: An application to optimize dehydration operations of selected agricultural crops. *LWT-Food Sci. Technol.* **2002**, *35*, 584–592. [CrossRef]
2. Sengo, I.; Gominho, J.; d'Orey, L.; Martins, M.; d'Almeida-Duarte, E.; Pereira, H.; Ferreira-Dias, S. Response surface modeling and optimization of biodiesel production from Cynara cardunculus oil. *Eur. J. Lipid Sci. Technol.* **2010**, *112*, 310–320. [CrossRef]
3. Silva, G.F.; Castro, M.S.; Silva, J.S.; Mendes, J.S.; Ferreira, A.L.O. Simulation and optimization of biodiesel production by soybean oil transesterification in nonideal continuous stirred-tank reactor. *Int. J. Chem. React. Eng.* **2010**, *8*. [CrossRef]
4. Tiwari, A.K.; Kumar, A.; Raheman, H. Biodiesel production from jatropha oil (Jatropha curcas) with high free fatty acids: An optimized process. *Biomass Bioenergy* **2007**, *31*, 569–575. [CrossRef]
5. Vicente, G.; Coteron, A.; Martinez, M.; Aracil, J. Application of the factorial design of experiments and response surface methodology to optimize biodiesel production. *Ind. Crops Prod.* **1998**, *8*, 29–35. [CrossRef]
6. Vicente, G.; Martınez, M.; Aracil, J. Integrated Biodiesel Production: A Comparison of Different Homogeneous Catalysts Systems. *Bioresour. Technol.* **2004**, *92*, 297–305. [CrossRef] [PubMed]
7. Dalvi, V.B.; Tiwari, K.N.; Pawade, M.N.; Phirke, P.S. Response surface analysis of tomato production under microirrigation. *Agric. Water Manag.* **1999**, *41*, 11–19. [CrossRef]
8. Photinam, R.; Moongngarm, A.; Paseephol, T. Process optimization to increase resistant starch in vermicelli prepared from mung bean and cowpea starch. *Emir. J. Food Agric.* **2016**, *28*, 449–458. [CrossRef]
9. Formato, A.; Ianniello, D.; Villecco, F.; Lenza, T.L.L.; Guida, D. Design Optimization of the Plough Working Surface by Computerized Mathematical Model. *Emir. J. Food Agric.* **2017**, *29*, 36–44. [CrossRef]
10. Muthuvelayudham, R.; Viruthagiri, T. Application of central composite design based response surface methodology in parameter optimization and on cellulase production using agricultural waste. *Int. J. Chem. Biol. Eng.* **2010**, *3*, 97–104.
11. Pishgar-Komleh, S.H.; Keyhani, A.; Mostofi-Sarkari, M.R.; Jafari, A. Application of response surface methodology for optimization of picker-husker harvesting losses in corn seed. *Iran. J. Energy Environ.* **2012**, *3*, 134–142. [CrossRef]
12. Alexandris, S.; Kerkides, P. New empirical formula for hourly estimations of reference evapotranspiration. *Agric. Water Manag.* **2003**, *60*, 181–198. [CrossRef]
13. Alexandris, S.; Kerkides, P.; Liakatas, A. Daily reference evapotranspiration estimates by the "Copais" approach. *Agric. Water Manag.* **2006**, *82*, 371–386. [CrossRef]
14. Poss, J.A.; Russell, W.B.; Shouse, P.J.; Austin, R.S.; Grattan, S.R.; Grieve, C.M.; Lieth, J.H.; Zeng, L. A volumetric lysimeter system (VLS): An alternative to weighing lysimeters for plant-water relations studies. *Comput. Electron. Agric.* **2004**, *43*, 55–68. [CrossRef]
15. Meulenberg, C.J.W.; Vijverberg, H.P. Empirical relations predicting human and rat tissue: Air partition coefficients of volatile organic compounds. *Toxicol. Appl. Pharmacol.* **2000**, *165*, 206–216. [CrossRef] [PubMed]
16. Wicki, A.; Parlow, E. Multiple Regression Analysis for Unmixing of Surface Temperature Data in an Urban Environment. *Remote Sens.* **2017**, *9*, 684. [CrossRef]
17. Pereira, L.S.; Teodoro, P.R.; Rodrigues, P.N.; Teixeira, J.L. Irrigation scheduling simulation: The model ISAREG. In *Tools for Drought Mitigation in Mediterranean Regions*; Rossi, G., Cancelliere, A., Pereira, L.S., Oweis, T., Shatanawi, M., Zairi, A., Eds.; Kluwer: Dordrecht, The Netherlands, 2003; pp. 161–180.
18. Allen, R.G.; Pereira, L.S.; Raes, D.; Smith, M. *Crop Evapotranspiration, FAO Irrigation and Drainage Paper No 56*; FAO: Rome, Italy, 1998; p. 327.

19. Fox, M.S. An organizational view of distributed systems. *IEEE Trans. Syst. Man Cybern.* **1981**, *11*, 70–80. [CrossRef]
20. Willmott, C.J. Some comments on the evaluation of model performance. *Bull. Am. Meteorol. Soc.* **1982**, *63*, 1309–1313. [CrossRef]

energies

MDPI

Article

Decision-Making Process for Photovoltaic Solar Energy Sector Development using Fuzzy Cognitive Map Technique

Konstantinos Papageorgiou [1,*], Gustavo Carvalho [2], Elpiniki I. Papageorgiou [3,4], Dionysis Bochtis [4] and George Stamoulis [1]

1 Department of Computer Science, University of Thessaly, Lamia 35100, Greece; gs@uth.gr
2 Faculty of Economics, Business Administration and Accounting, University of Sao Paulo (USP), São Paulo 05508-010, Brazil; gustavocarvalho@usp.br
3 Faculty of Technology, University of Thessaly, Geopolis Campus, Larissa 41500, Greece; elpinikipapageorgiou@uth.gr
4 Institute for Bio-economy and Agri-technology (iBO), Center for Research and Technology – Hellas (CERTH), Thessaloniki 57001, Greece; d.bochtis@certh.gr
* Correspondence: konpapageorgiou@uth.gr

Received: 25 February 2020; Accepted: 16 March 2020; Published: 19 March 2020

Abstract: Photovoltaic Solar Energy (PSE) sector has sparked great interest from governments over the last decade towards diminution of world's dependency to fossil fuels, greenhouse gas emissions reduction and global warming mitigation. Photovoltaic solar energy also holds a significant role in the transition to sustainable energy systems. These systems and their optimal exploitation require an effective supply chain management system, such as design of the network, collection, storage, or transportation of this energy resource, without disregarding a country's certain socio-economic and political conditions. In Brazil, the adoption of photovoltaic solar energy has been motivated not only by the energy matrix diversification but also from the shortages, problems, and barriers that the Brazilian energy sector has faced, lately. However, PSE development is affected by various factors with high uncertainty, such as political, social, economic, and environmental, that include critical operational sustainability issues. Thus, an elaborate modelling of energy management and a well-structured decision support process are needed to enhance the performance efficiency of Brazilian PSE supply chain management. This study focuses on the investigation of certain factors and their influence on the development of the Brazilian PSE with the help of Fuzzy Cognitive Maps. Fuzzy Cognitive Map is an established methodology for scenario analysis and management in diverse domains, inheriting the advancements of fuzzy logic and neural networks. In this context, a semi-quantitative model was designed with the help of various stakeholders from the specific energy domain and three plausible scenarios were conducted in order to support a decision-making process on PSE sector development and the country's economic potential. The outcome of this analysis reveals that the development of the PSE sector in Brazil is mainly affected by economic and political factors.

Keywords: Fuzzy Cognitive Maps; photovoltaic solar energy; scenario analysis; decision-support; energy management

1. Introduction

Fuzzy cognitive maps (FCMs) were introduced by Kosko (1986) as a methodology for modelling complex, dynamic systems [1]. In terms of structure, FCM is a directed weighted graph with feedback loops, which consists of nodes (concepts) and strength connections (relationships) among nodes [2]. An FCM models the system in the form of a graph, introducing causal relations between concepts and represents the system's behavior using the accumulated knowledge, perceptions, experiences or beliefs from experts, stakeholders or both [3,4], usually expressed in natural language.

Over the last years, FCMs have attained great attention and popularity since they are able to implement modelling, analyzing and simulating tasks, as well as test the influence of parameters and predict the behavior of the examined system [2]. Furthermore, they include learning capabilities and characteristics as they can learn from historical data and previous knowledge to overcome the subjectivity of experts' opinions [5–7]. In other words, it is a semi-quantitative method that uses qualitative knowledge from experts and stakeholders to form FCM-based models and further increase system understanding and reduce uncertainties.

The task of simulation is incorporated by FCMs and can help managers to increase system understanding and reduce uncertainties about how the studied phenomenon may respond to changes in the main system drivers [8].

Among other fields, FCMs are also used for scenario planning even if they were not initially used for this purpose [9]. Kok [10], Van Vliet et al. [11] and Jetter and Schweinfort [12] have introduced FCM-based scenarios to improve the quality of scenarios and increase the robustness of the scenario development step.

Over the last two decades, renewable energy has been considered a vital candidate in global power demand [13], and thus, its exploitation needed to be well explored and studied. However, due to several limitations, such as environmental problems and because of the high uncertainty that these kinds of systems present, the renewable energy sector is considered highly non-linear and complex. In this context, Fuzzy Cognitive Map is a method that can illustrate the uncertainty causality [14] and becomes a powerful tool to model such complex systems that involve many factors [2]. Due to this fact, FCM has been recently deployed in a renewable energy domain with great perspectives.

Taking one step further, FCMs have recently implemented management and decision-making tasks in the Energy sector [9,15]. In particular, in [15], an FCM approach was applied to predict and analyze the consequences of Brexit for the Energy-Water-Food (EWF) demand, examining different Brexit scenarios. FCMs can deal with complex problems, vast domains of inherent uncertainties [16] and processes like decision making that are based on human reasoning process [2,17]. In [18] and [19], special attention has been given to FCMs for management, decision-making and prediction purposes in the energy context. Specifically, the authors in [18] have proposed an algorithm based on Artificial Neural Networks (ANN) to create a model for monthly gasoline consumption estimation. The produced results could help officials and producers for better planning and marketing as well as optimizing gasoline production. An FCM has been proposed in [20] as a powerful inference tool to model the complex system of the electricity market. In this context, the authors describe the structure of FCMs and examine the way they are used to model and simulate competitive trading markets. Thus, certain approaches will be selected to further increase the understanding of how such markets behave in different circumstances. A Fuzzy logic-based framework was investigated in [21] for anomaly detection in energy consumption data where uncertainties inherent to forecasting, have been tackled. Regarding the renewable energy domain, a socio-economic analysis was conducted in [22], in which FCMs were employed to identify potential consequences over the mix of fossil fuel and renewable energy power sectors, as well as how certain policies that regard the energy domain can affect a country's economic growth. We also need to report on [23], in which an FCM methodology was applied for assessing energy efficiency strategies and electricity demand prediction, whereas Nikas et al., dealt with policy-making in the Renewable Energy sector, employing a stakeholder-driven approach based on FCMs [24]. Alipour et al., in a more recent study, explored the FCM capabilities on identifying and analyzing unpredictable factors that have a direct impact on solar photovoltaics in an uncertain socio-economic, political and technological environment [25]. It is worth mentioning that this is the first research study that implements FCM-based scenario analysis in the renewable energy field.

In order to tackle the vast amount of data, various parameters, modelling, learning, and simulation tasks of complex systems, researchers need a simple but powerful software tool that could help them in this direction. As found on the related literature, several tries occurred by researchers for developing a software product with simple or advanced functions that could perform either designing and analyzing tasks or even more complex functions like learning and simulation on FCM-based systems. Reviewing the existing literature, tools like Mental Modeler [26], FCMapper [27], ESQAPE [23], FCM Expert [28], FCM-Analyst [29], JFCM [30], and FCM Designer [31] can offer various capabilities and have their own unique specifications. However, MentalModeler [26] is the only web-based tool for scenario analysis and environmental modeling. This tool provides an opportunity to incorporate different types of knowledge into environmental decision-making, define hypotheses to be tested, and run scenarios to determine perceived outcomes of proposed policies.

However, a gap is noticed in the relevant literature, regarding the existence of a more user-friendly tool that could provide extra capabilities for aggregating FCM models and learning them using supervised and unsupervised algorithms. Considering this gap, the team of Papageorgiou E. has developed a new web-based tool, called FCMwizard [25], that incorporates these functionalities for helping research community to freely experiment with them and exploit all the capabilities offered. Some preliminary work with the use of the FCMWizard tool has been recently made in the environmental domain, dealing with decision making problems such as (i) the economic sustainability of poor women in rural areas [32] and (ii) climate change adaptations [33].

In this work, the FCM technique is applied to perform a decision-making process for PSE sector development, providing at the same time the researchers and the FCM community, wth an efficient simulation tool called FCMWizard, to conduct policy making simulations in their own case studies. In particular, the authors explore relevant FCM-based scenarios for the Brazilian Photovoltaic Solar Energy Sector (PSE) employing the FCMWizard tool. The PSE arises as one of the most promising renewable energy sources to replace fossil fuels [34], since climate and environmental changes, described by the scientific community [35], are increasing continuously. This happens as a result of the increasing energy demand that stems from technological progress and other advancements in human development. Scenario planning has proven to be the appropriate technique for institutions that are immersed in this PSE sector and need to make complex decisions, taking into account various uncertainties and risks, such as socio-political, environmental and technological.

The following tasks emphasize the novelty of this research work:

(i) to present the decision-making capabilities of the new FCMWizard tool for modelling stakeholders' perception in the complex sector of Renewable Energy, and
(ii) to develop and investigate various scenarios using an FCM-based simulation process, implemented by the new FCMWizard tool, regarding the development of PSE in Brazil.

The structure of the present paper is as follows: Section 2 briefly describes FCMs, while Section 3 presents the problem of PSE development in Brazil and the building process of the model. In Section 4, FCMWizard tool is adequately presented and its decision-making functionalities are reported accordingly. The scenarios development and analysis regarding the Brazilian PSE with the use of the FCMs are provided in Section 5. The discussion of the results is laid in the same section. The last section is devoted to the conclusions of this study.

2. Fuzzy Cognitive Maps

FCMs are fuzzy structures that combine fuzzy logic and Neural Networks, which incorporate knowledge acquisition and indicate the causality among concepts of complex systems that are characterized by uncertainty. FCMs were first introduced by Kosko [1] and since then they have been used to represent human knowledge and reasoning, while they serve as a method for modelling dynamic decision systems. Furthermore, learning techniques were later included to train FCMs and identify appropriate weights for its interconnections.

FCMs as we know them, are fuzzy signed directed graphs with feedback. They consist of concepts that are represented in the graph by nodes. Concepts are interconnected with directed arcs that are labelled with fuzzy values in the interval [0, 1] or [−1, +1] and show the strength of impact between the concepts [36]. Typically, an FCM with n nodes can be represented in the form of an adjacency matrix W_{nxn}, where each element w_{ij} of the matrix takes values within [−1, +1].

The causal relationships (weights) between concepts have three possible types.

- Positive causality if $w_{ij} > 0$, where Ci increases C_j
- Negative causality if $w_{ij} < 0$, where Ci decreases C_j
- No causality between C_i and C_j if $w_{ij} = 0$.

The FCM converges to an equilibrium point using one of the following calculation rules, (1)–(3), when values X_i have been assigned to the concepts C_i [37]:

$$\text{Kosko}: X_i^{(\kappa+1)} = f\left(\sum_{\substack{j=1 \\ j \neq i}}^{n} w_{ji} \times X_j^{\kappa}\right) \tag{1}$$

$$\text{Modified Kosko}: X_i^{(\kappa+1)} = f\left(X_i(t) + \sum_{\substack{j=1 \\ j \neq i}}^{n} X_j(t) \cdot w_{ji}\right) \tag{2}$$

$$\text{Rescale}: X_i^{(\kappa+1)} = f\left(\left(2 \times X_i^{\kappa} - 1\right) + \sum_{j=1,\, j\neq i}^{n} w_{ji} \times \left(2 \times X_j^{\kappa} - 1\right)\right) \tag{3}$$

where $X_i^{(\kappa+1)}$ is the value of concept C_i at simulation step $\kappa + 1$, X_j^{κ} is the value of concept C_j at the simulation step κ, w_{ij} is the weight of the interconnection between concept C_i and concept C_j, κ (Greek letter Kappa) is the interaction index and every simulation step, and f(·) is the threshold (activation) function. The function is selected to retain the values within the range [0, 1] or [−1, 1]. Equation (2) is the original Kosko activation function (1) that takes past values (or history) of a concept into account [38]. Generally, there are four most used transformation functions (4)–(7): (a) bivalent, (b) trivalent, (c) sigmoid, and (d) hyperbolic tangent.

$$\text{Bivalent}: f(x) = \begin{cases} 1, & x < 0 \\ 0, & x \geq 0 \end{cases} \tag{4}$$

$$\text{Trivalent}: f(x) = \begin{cases} 1, x > 0 \\ 0, x = 0 \\ -1, x < 0 \end{cases} \tag{5}$$

$$\text{Sigmoid (linear)}: f(x) = \frac{1}{1 + e^{-\lambda x}} \tag{6}$$

$$\text{Hyperbolic tan gent}: f(x) = \tanh(\lambda x) \tag{7}$$

where λ is a real positive number ($\lambda > 0$), which determines the steepness of the continuous function f and x is the value X_i^{κ} on the equilibrium point. At each step, the value X_i of a concept is influenced by the values of concepts connected to it and it is updated according to the inference rule. This process continues until the system converges, which indicates that the difference between two subsequent value of the outputs values must be equal or lower to ε (epsilon, $\varepsilon = 0.001$).

3. PSE in Brazil

3.1. Problem Statement

Environmental degradation caused by deforestation, gas emission, and environmental pollution has caused countries to rethink their electric systems. As a solution to the problems caused by the unsustainable exploration of fossil fuels, renewable energies have become the focus of attention of a broad range of agents. However, in Brazil, unlike other countries, the investment in photovoltaic solar energy was driven by different reasons, such the increase in the cost of energy produced in thermoelectric plants and the emission of greenhouse gases by burning of fossil fuels [39,40].

Among other solutions, PSE shows up as a viable and necessary source of electrical energy. Given that it is located mostly in the intertropical region and has excellent potential for solar energy utilization throughout the year, the Brazilian solar radiation is higher than the European for almost its entire territory [41]. Based on Renewable Energy Country Attractiveness Index [42], Brazil is eighth in the raking of attractiveness when speaking of photovoltaic solar energy. In 2014, Brazil had its total installed to power up to 15 MW, and in 2015, the same number surpassed 32 MW. Statistics of the Mines and Energy Ministry [43] indicated that by 2018 Brazil should be between the 20 countries with the most prominent generation of solar energy.

On the other hand, the high cost of acquisition of photovoltaic and their low conversion efficiency from solar irradiation to electrical energy are the main impediments to the large-scale diffusion of these systems [44,45]. As a solution, the government can understand the dynamic of PSE development from other countries and propose policies that improve the use of solar energy in an urban environment [46]. Given that the future of the PES depends on several issues, it needs to be planned and controlled. For this purpose, scenario planning is a fundamental tool [47].

3.2. Model Development

In order to provide a quasi-quantitative model for scenario planning, regarding the studied problem of PSE development in Brazil, the FCM development process was conducted in six steps. Figure 1 offers a visual representation of these steps for better understanding of the procedure and for the convenience of the readers.

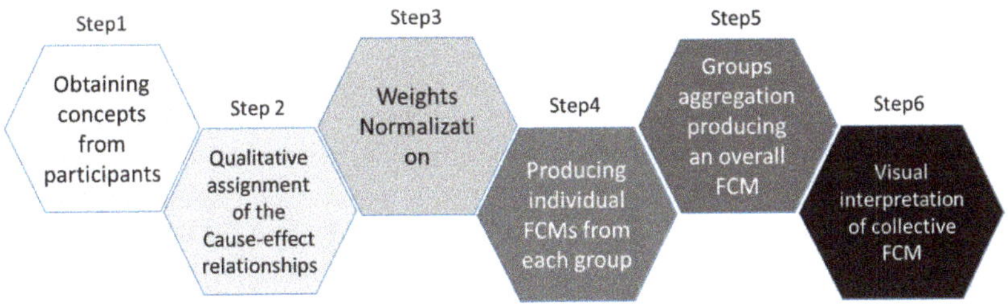

Figure 1. The steps of FCM development process.

Step 1: Obtaining concepts from participants.

The first step deals with the determination of the most essential independent variables that define the examined problem and affect the dependent variable. Participants are asked to determine the concepts of the system but only the most important in order to avoid designing a system with a vast number of variables that would be difficult to study.

Step 2: Qualitative assignment of cause-effect relationships between concepts.

In the second step, participants are asked to assign values on the scale of 1–10 and to determine whether there is a positive or negative cause-effect relationship for the weighted interconnections among the depended and most significant variables that constitute the FCM model. Ten (10) denotes the highest strength and one (1) the lowest. The sign + (plus) denotes a positive influence, whereas the sign – (minus) denotes a negative influence that one node has to another.

Step 3: Weights Normalization (coding into an adjacency matrix).

Weights given to each link were then normalized between 0 and 1 (as described in Table 1), considering positive and negative values for coding into the adjacency matrix [48]. So, we substitute the qualitative values assigned by the stakeholders for expressing the degree of influence with the corresponding quantitative values, in order to define the weight matrix of each stakeholder, as presented in Table 1. For example, the qualitative value 10 (that denotes the highest strength) is substituted by the quantitative weight value 1.

Table 1. Normalizing Qualitative (linguistic) values to Quantitative values for the weighted interconnections.

Qualitative Opinion (Linguistic Values)	Weight Value (Quantitative)	Qualitative Opinion (Linguistic Values)	Weight Value (Quantitative)
1 - Extremely low	0.1	6 - Medium high	0.6
2 - Very low	0.2	7 - High	0.7
3 - Low	0.3	8 - Very High	0.8
4 - Medium low	0.4	9 - Extremely high	0.9
5 - Medium	0.5	10 - Highest strength	1

Step 4: Producing individual FCMs from each group.

In this step, every group of the participants was asked to construct an individual FCM by defining the primary variables and determining the weights values of all causal relationships. This process also includes the identification of the decision concept, which is a dependent variable for the problem under investigation. As we aim to analyze the Brazilian Solar Photovoltaic Energy Sector, the dependent variable was set to be the "Development of PSE in Brazil". The procedure that was followed, is described below in detail, as the authors want to highlight the importance of all the steps taken for the success of this study.

Step 5: Groups Aggregation producing an Overall FCM.

In this step, all individually coded cognitive maps from the three groups were aggregated and an overall group FCM (Collective-FCM) was produced that includes all the concepts from all individual cognitive maps. This collective FCM represents the perception of all the stakeholders and is enriched with the knowledge of all stakeholders involved.

Step 6: Visualization of collective FCM.

After the aggregation process, that was based on the weighted average method, the Collective-FCM was analyzed using the FCMWizard software tool (version 1.0, E.I. Papageorgiou, Larissa, Greece). Since the tool includes modelling and visualization capabilities, a visual representation of the condensed FCM model was created by FCMWizard, which specifically illustrates the concepts and all the connections between them. The graphical representation of the collective FCM is presented in Section 4.1, where an overview of the FCMWizard tool is presented.

Following the steps above, the authors carried out the project with the help of researchers of the University of São Paulo (USP) and University of Brasília (UNB) specialized in the solar photovoltaic energy. They along with other specialized stakeholders (such as specialists from the National Electric Energy Agency (government), the Brazilian Solar Energy Association (professionals)) were the primary sources of data and information for the development of this research.

Interviews were conducted individually or with pairs of specialists from the National Electric Energy Agency (government), the World Wildlife Fund (NGO), Brazilian Solar Energy Association (professionals), and researchers from the University of São Paulo and Brasília (specialists). A workshop was also carried out with a group of eight graduate students of the Institute of Energy and Environment of the University of São Paulo, IEE/USP, to consolidate the FCMs.

Specifically, a pretest was first carried out with a potential consumer with a business background, through an individual interview, during which the dynamics of FCMs and the proposed method were explained. The respondent had trouble defining the primary variables (concepts) and establishing the exact weights (from −1 to 1) of the causal relations. After proper guidance and following the contents of Table 1, in the end, he managed to develop a potential FCM.

The second FCM was constructed after interviewing a specialist in SPE from the University of Brasília (UNB). He responded with great enthusiasm and a consensus was reached for 12 interrelated concepts, establishing at the same time the causal relations and respective weights.

In the next FCM, the proposed method was investigated with the help of a group of SPE specialists from IEE/USP. In a workshop with eight participants, the dynamics of FCMs and the proposed method were explained. These specialists identified 13 concepts and established their connections and weights. In this type of approach (workshop), the most prominent advantage came from the debate among the participants. Instead of creating individual FCMs, the authors suggested a collaborative process for scenario planning where the participants cooperated to produce a more productive and accurate knowledge in the form of a consolidated FCM. The weight value for each interconnection is calculated as the average of all values that each participant gave for the corresponding interconnection, taking into consideration Table 1. However, the FCM achieved by the group of specialists from USP, presented in Figure 2, was not a significant improvement regarding complexity and robustness in comparison with the others.

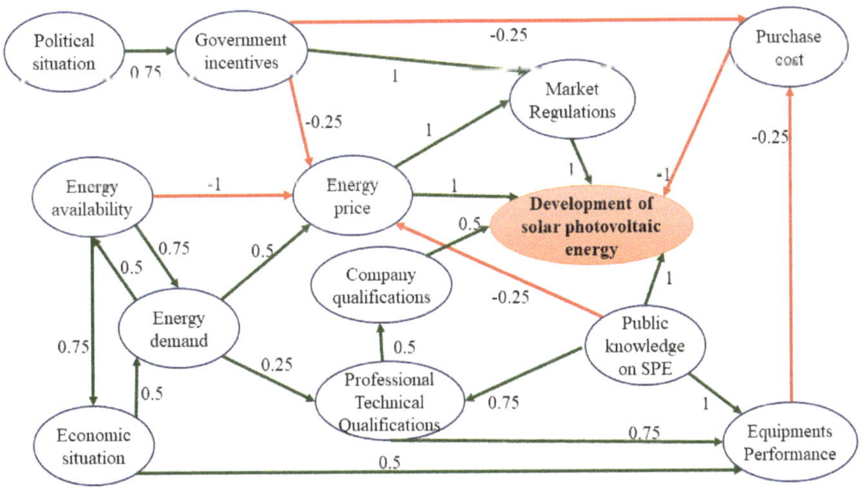

Figure 2. FCM from a group of the specialist of IEE/USP (average weight values).

The FCMs elaborated from the information of the specialists, and stakeholders were integrated into a single FCM. Table 2 presents the list of the concepts contained in this collective FCM, with a short description of them. They identified 29 concepts in total and established the connections and weights among these concepts. Also, they determined the 10 most central concepts (concepts in bold in Table 2) in terms of significance in the decision making process through scenario analysis. Moreover, the weights of the connections were equal to the average of the weights established by the specialists and stakeholders.

Table 2. List of concepts integrating the collective FCM.

Concept	ID	Description
Development of solar photovoltaic energy	**C1**	Possible evolution of the Photovoltaic Solar Energy Sector in Brazil
Potential generating jobs	C2	Refers to the capacity to create employment.
Market competition	C3	
Public knowledge of SPE	**C4**	Society's understanding and awareness of the functioning and role of PS energy in energy systems and everyday use.
Public opinion (ecological and social consciousness)	C5	Public awareness regarding the importance of environmental protection
Purchase cost	**C6**	The price of energy product or services that are available for purchase by suppliers or consumers.
Maintenance costs	C7	The expenses incurred to keep an item in good condition or good working order
Energy demand	**C8**	The quantity of energy that consumers wish to acquire for a set price in a market.
Energy availability	C9	The amount of energy from various energy recourses that is available for consumption
Decentralized energy creation	C10	Energy production facilities closer to the site of energy consumption that allow for more optimal use of renewable energy
Government incentives	**C11**	They are benefits granted by governments to foster the progress of the sector (tax exemptions, financing lines, demand creation).
Private sector involvement	**C12**	The participation of the private sector in energy projects of the government.
Geographical location	C13	Geographic location refers to a position of the country on Earth. Defined by longitude and latitude.
Energy dependence concerns	**C14**	Energy source dependence. Need to diversify the energy matrix.
ABNT (Brazilian Technical Standards)	C15	A private non-profit organization which is responsible for technical standards in Brazil, and intends to promote technological development in the country
Payback Period	**C16**	The length of time required for an investment to recover its initial outlay in terms of profits or savings.
Environmental pollution	C17	The addition of any substance or any form of energy to the environment at a rate faster than it can be dispersed, diluted or decomposed.
Energy price	**C18**	The value of the energy expressed in monetary terms.
National production equipment	C19	Production equipment that belongs to a country and its purpose is to produce goods of a wanted quality when provided with production resources of a required quality
Equipment performance	C20	Effectiveness of output of a given piece of equipment.
Companies' qualifications	C21	Certain requirements or certificates for specific occupational needs.
Professional technical qualifications	C22	Associated with how excellent the professionals' skills are. It means the attributes and characteristics of a worker.
Availability of alternative energies	C23	The amount of energy from alternative energy recourses that is available for consumption
Market regulations	C24	Government regulations to ensure that markets function effectively and efficiently.
Economic situation	**C25**	It is the capacity of an economy to produce goods and services. In Brazil, the most common proxy used is a Gross domestic product (GDP).
Global treaties	C26	A formal agreement between two or more independent governments that impact the energy sector (Kyoto Protocol, Montreal Protocol).
Taxation	C27	Taxes account for a significant share of the final prices consumers pay for energy and can have a significant impact on consumption and investment patterns.
Political situation	C28	The way power is achieved and used in a country
Technological maturity	C29	Technological progress or growth

In this context, a 29-nodes FCM model was produced from the experts/stakeholders and its adjacency matrix (see Figure 3) was created accordingly, including the causal relationships between the concepts and their weight.

0	0	0	0	0	0	0	0	0	0	0	0	0	0	0	0	0	0	0	0	0	0	0	0	0	0	0	0	0
0	0	0	0	0	0	0	0	0	0	0	0	0	0	0	0	0	0	0	0	0	0	0	0	0.25	0	0	0	0
0.5	0	0	0	0	0.5	0	0	0	0	0	0	0	0	0	-0.25	0	0	0	0	0	0	0	0	0	0	0	0	0
0.63	0	0	0	0.75	0	0	0	0	0	0	0	0	0	0	0	0	-0.25	0	1	0	0.75	0	0	0	0	0	0	0
0.38	0	0	0	0	0	0	0	0	0	0	0.25	0	0	0	0	-0.5	0	0	0	0	0	0	0	0	0	0	0	0
-0.92	0	0	0	0	0	0	0	0	0	0	0	0	0	0	0.83	0	0	0	0	0	0	0	0	0	0	0	0	0
0	0	0	0	0	0	0	0	0	0	0	0	0	0	0	0.25	0	0	0	0	0	0	0	0	0	0	0	0	0
0.5	0	0	0	0	0	0	0	0.5	0	0	0	0	0.88	0	0	0	0.25	0	0	0	0.25	0	0	0	0	0	0	0
0	0	0	0	0	0	0	0.75	0	0	0	0	0	0.75	0	0	0	-1	0	0	0	0	0	0	0.75	0	0	0	0
1	0	0	0	0	0	0	0	0	0	0	0	0	0	0	0	0	0	0	0	0	0	0	0	0	0	0	0	0
0.67	0.75	0	0	0	-0.63	0	0	0	0	0	0.63	0	0	0	-0.5	0	-0.25	-0.5	0	0	0	0	1	0	0	0	0	0.5
0.5	0	0.25	0	0	0	0	0	0	0	0	0	0	0	0	0	0	-0.25	1	0	0	0	0	0	0	0	0	0	0.5
0.38	0	0	0	0	0	0	0	0	0	0	0	0	0	0	0	0	0	0	0	0	0	0	0	0	0	0	0	0
0.75	0	0.5	0	0	0	0	0	0	1	0	0	0	0	0	0	0	0	0	0	0	0	0	0.5	0	0	0	0	0
0.5	0	0	0	0	0	0	0	0	0	0	0	0	0	0	0	0	0	0	0	0	0	0	0	0	0	0	0	0
-0.63	0	0	0	0	0	0	0	0	0	0	0	-0.25	0	0	0	0	0	0	0	0	0	0	0	0	0	0	0	0
0	0	0	0	0	0	0	0	0	0	0	0	0	0.5	0	0	0	0	0	0	0	0	0	0	0	0.75	0	0	0
0.75	0	0	0.5	0	0	0	0	0	0	0	0	0	0	0	-0.8	0	0	0	0	0	0	0	1	-0.5	0	0	0	0
0.5	0	0	0	0	0	0	0	0	0	0	0	0	0	0	0	0	0	0	1	0	0	0	0	0	0	0	0	0
0.5	0	0	0	0	-0.25	0	0	0	0	0	0	0	0	0	0	0	0	0	0	0	0	0	0	0	0	0	0	0
0.5	0	0	0	0	0	0	0	0	0	0	0	0	0	0	0	0	0	0	0.75	0	0	0	0	0	0	0	0	0
0	0	0	0	0	0	0	0	0	0	0	0	0	0	0	0	0	0	0	0	0.5	0	0	0	0	0	0	0	0
0	0	0	0	0	0	0	0	0	0	0	0	0	0.25	0	0	0	0	0	0	0	0	0	0	0	0	0	0	0
0.75	0	0	0	0	0	0	0	0	0	0.25	0	0	0	0	-0.5	0	0	0.25	0	0	0	0	0	0	0	0	0	0
1	0.25	0	0	0	0	0	0.5	0	0	0	0.5	0	0	0	0	0	-0.25	0	0.5	0	0	0	0	0	0	0	0	0
0.25	0	0	0	0	0	0	0	0	0	0	0	0	0	0	0	0	0	0	0	0	0	0	0	0	0	0	0	0
0	0	0	0	0	0.25	0	0	0	0	0	0	0	0	0	0.5	0	0	-0.5	0	0	0	0	0	0	0	0	0	0
0.25	0	0	0	0	0	0	0	0	0	0.63	0	0	0	0	0	0	0	0	0	0	0	0	0	1	0	0	0	0
0	0	0	0	0	-1	0	0	0	0	0	0	0	0	0	0	0	0	0	1	0	0	0	0	0	0	0	0	0

Figure 3. Adjacency matrix.

4. Overview of the FCMwizard Tool

This section provides an overview of the FCMWizard tool in which the functionalities of this software are presented. In order to illustrate the capabilities of this tool on modelling and simulating a dynamic and complex system, the authors use the case study of the renewable energy sector of Brazil. Among the functionalities that site on the main menu (see Figure 4), there are three main modes for constructing an FCM, namely: (i) Expert mode wherein experts and/or stakeholders can construct manually a FCM for a real problem; (ii) Data-based Learning mode wherein the FCM model is constructed automatically using given data; and (iii) Merge mode, where different individual FCMs can be combined to provide an augmented and collective FCM model in order to generate aggregated system complexities.

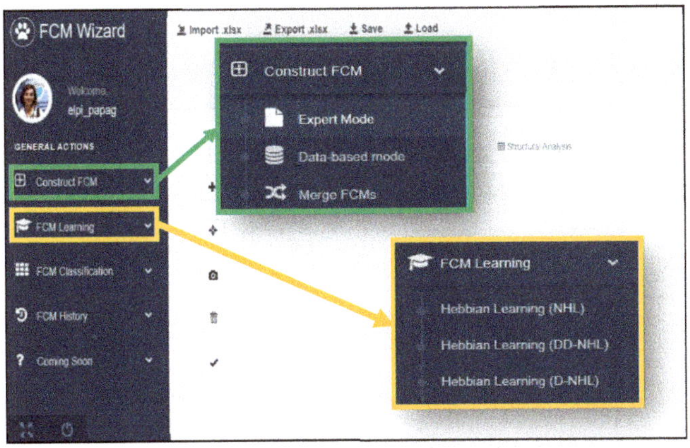

Figure 4. FCMWizard Main menu, with FCM construction and FCM Learning modes.

FCMWizard offers a Main menu with 2 main functionalities: FCM construction and FCM Learning. FCM Learning includes three learning modes (see Figure 4): (i) Hebbian Learning (NHL): weights are fine-tuned by data assuming that all the concepts are synchronously triggered at each iteration and they synchronously change their values. Constraints on concepts and weights can be considered for updating weighted connections. (ii) Data-driven Hebbian Learning (DD-NHL): an extended NHL method through historical data for the input concepts to fine-tune weights, and (iii) Differential Hebbian Learning (D-NHL): This algorithm correlates the changes in the concepts to modify their weights [2].

Next, the authors will illustrate how the FCM Wizard tool can be used to construct an FCM model and visualize its results for the given problem of PSE in Brazil. Moreover, the structural analysis, the Hebbian-based Learning and the scenario analysis will be further described, along with relevant figures and tables, produced by the tool.

4.1. Expert-Based Construction/Modelling

Without a deeper knowledge of the mathematical foundations of the methodology and regardless of the type of application domain, users can create new concepts and causal relationships between them, using influence weights, to design a new model. The "Expert Mode" under the main menu "Construct FCM", provides users with the ability to easily design an FCM model by adding new nodes when the "+" button in the "Model Design" tab is selected. Then, new directed lines can be drawn representing the causal relationships between nodes, in which a green direct relationship denotes a positive weight and a red inverse relationship denotes a negative weight.

So, the FCM model that was designed by the stakeholders can be easily constructed, as illustrated in Figure 5, using the FCMWizard tool, following the process described above. This consensus FCM model comprises the concepts in the form of nodes (C1, C2, ... , C29) and directed lines (green or red) with the corresponding weight (positive or negative). Moreover, this FCM model can be directly designed by the FCMWizard, after properly importing the adjacency matrix (see Figure 3) in the tool, as provided by the Stakeholders. The weight matrix, as illustrated in the tab "weight matrix" of the tool, is presented in Figure 6. Each number of this weight matrix represents the causal relationship between the concepts that are sited in the corresponding row and column of this matrix.

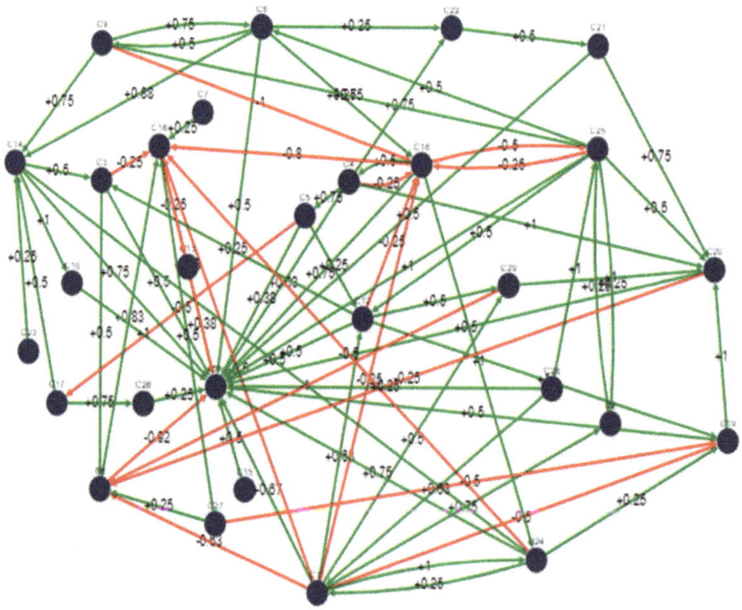

Figure 5. Consensus FCM model for PSE in Brazil designed in FCMwizard tool.

Model Design | Weight Matrix | Scenario Analysis | Structural Analysis

Weight Matrix

	C1	C2	C3	C4	C5	C6	C7	C8	C9	C10	C11	C12	C13	C14	C15	C16	C17	C18	C19	C20	C21	C22	C23	C24	C25	C26	C27	C28	C29
C1	0	0	0	0	0	0	0	0	0	0	0	0	0	0	0	0	0	0	0	0	0	0	0	0	0	0	0	0	0
C2	0	0	0	0	0	0	0	0	0	0	0	0	0	0	0	0	0	0	0	0	0	0	0	0	0.25	0	0	0	0
C3	0.5	0	0	0	0	0.5	0	0	0	0	0	0	0	0	0	-0.25	0	0	0	0	0	0	0	0	0	0	0	0	0
C4	0.63	0	0	0	0.75	0	0	0	0	0	0	0	0	0	0	0	0	-0.25	0	1	0	0.75	0	0	0	0	0	0	0
C5	0.38	0	0	0	0	0	0	0	0	0	0	0.25	0	0	0	0	-0.5	0	0	0	0	0	0	0	0	0	0	0	0
C6	-0.92	0	0	0	0	0	0	0	0	0	0	0	0	0	0	0.83	0	0	0	0	0	0	0	0	0	0	0	0	0
C7	0	0	0	0	0	0	0	0	0	0	0	0	0	0	0	0.25	0	0	0	0	0	0	0	0	0	0	0	0	0
C8	0.5	0	0	0	0	0	0	0	0.5	0	0	0	0	0.83	0	0	0	0.25	0	0	0	0.25	0	0	0	0	0	0	0
C9	0	0	0	0	0	0	0	0.75	0	0	0	0	0	0.75	0	0	0	-1	0	0	0	0	0	0	0.75	0	0	0	0
C10	1	0	0	0	0	0	0	0	0	0	0	0	0	0	0	0	0	0	0	0	0	0	0	0	0	0	0	0	0
C11	0.67	0.75	0	0	0	-0.83	0	0	0	0	0	0.63	0	0	0	-0.5	0	-0.25	-0.5	0	0	0	0	1	0	0	0	0	0.5
C12	0.5	0	0.25	0	0	0	0	0	0	0	0	0	0	0	0	0	0	-0.25	1	0	0	0	0	0	0	0	0	0	0.5
C13	0.38	0	0	0	0	0	0	0	0	0	0	0	0	0	0	0	0	0	0	0	0	0	0	0	0	0	0	0	0
C14	0.75	0	0.5	0	0	0	0	0	0	1	0	0	0	0	0	0	0	0	0	0	0	0	0	0.5	0	0	0	0	0
C15	0.5	0	0	0	0	0	0	0	0	0	0	0	0	0	0	0	0	0	0	0	0	0	0	0	0	0	0	0	0
C16	-0.83	0	0	0	0	0	0	0	0	0	0	0	-0.25	0	0	0	0	0	0	0	0	0	0	0	0	0	0	0	0
C17	0	0	0	0	0	0	0	0	0	0	0	0	0	0.5	0	0	0	0	0	0	0	0	0	0	0	0.75	0	0	0
C18	0.75	0	0	0.5	0	0	0	0	0	0	0	0	0	0	0	-0.8	0	0	0	0	0	0	0	1	-0.5	0	0	0	0
C19	0.5	0	0	0	0	0	0	0	0	0	0	0	0	0	0	0	0	0	0	1	0	0	0	0	0	0	0	0	0
C20	0.5	0	0	0	0	-0.25	0	0	0	0	0	0	0	0	0	0	0	0	0	0	0	0	0	0	0	0	0	0	0
C21	0.5	0	0	0	0	0	0	0	0	0	0	0	0	0	0	0	0	0	0	0.75	0	0	0	0	0	0	0	0	0
C22	0	0	0	0	0	0	0	0	0	0	0	0	0	0	0	0	0	0	0	0	0.5	[illegible]	[illegible]	[illegible]	[illegible]	[illegible]	[illegible]	[illegible]	0
C23	0	0	0	0	0	0	0	0	0	0	0	0	0	0.25	0	0	0	0	0	0	0	0	0	0	0	0	0	0	0
C24	0.75	0	0	0	0	0	0	0	0	0	0.25	0	0	[illegible]	[illegible]	[illegible]	0	[illegible]	0.25	0	0	0	0	0	0	0	0	0	0
C25	1	0.25	0	0	0	0	0	0.5	0	0	0	0.5	0	0	0	0	0	-0.25	0	0.5	0	0	0	0	0	0	0	0	0
C26	0.25	0	0	0	0	0	0	0	0	0	0	0	0	0	0	0	0	0	0	0	0	0	0	0	0	0	0	0	0
C27	0	0	0	0	0	0.25	0	0	0	0	0	0	0	0	0	0.5	0	0	-0.5	0	0	0	0	0	0	0	0	0	0
C28	0.25	0	0	0	0	0	0	0	0	0	0.63	0	0	0	0	0	0	0	0	0	0	0	0	0	1	0	0	0	0
C29	0	0	0	0	0	-1	[illegible]	[illegible]	[illegible]	[illegible]	[illegible]	[illegible]	0	0	0	0	0	0	0	1	0	0	0	0	0	0	0	0	0

Figure 6. Screenshot of weight matrix of the final consensus FCM model produced in FCMWizard.

4.2. *Structural Analysis*

This software module encapsulates the ability to calculate various graph theory indices like the total number of concepts, number of connections, connection-to-concept ratio, outdegree, indegree, type of variables (receiver, transmitter or ordinary), degree centrality, betweenness centrality, closeness centrality, complexity ratio, density, and hierarchy index [33,49]. In Figure 7, a snapshot of the first eight concepts' graph indices is illustrated. Centrality is considered a significant index, so that a concept with a high degree of centrality has an important role in the cognitive map [8,49]. Centrality is calculated by the sum of the corresponding absolute indegree and outdegree causal weights [50]. Table 3 gathers the two most important indices of centrality (Degree and Betweenness) for all 29 concepts, as calculated by the tool. As observed from Figure 7 and Table 3, the concepts C1, C4, C6, C8, C11, C14, C16, C18, C24, and C25 present a higher degree of centrality than the rest of the concepts, so they can have a significant role on the examined energy system.

Complexity, density and hierarchy index too, are among significant structural indices for analyzing the graphical structure of FCMs.

From the constructed FCM model (Figure 5), it emerges that concepts with the most substantial number of connections are: Public knowledge on SPE (C4), Public opinion (C5), Energy demand (C8), Government incentives (C11), Private sector involvement (C12), Environmental pollution (C17), Energy price (C18), and Economic situation (C25). These key concepts are also directly and strongly connected to the objective concept C1 (development of solar photovoltaic energy). Hence, in terms of their degree of centrality and connectedness (see Figure 7 and Table 3), concepts C4, C5, C11, C17, and C18 were chosen by the researchers to become the base of the scenarios since they can strongly affect the behavior of the system.

(Nodes) Graph Indices

	C1	C2	C3	C4	C5	C6	C7	C8
Outdegree	0.0	0.3	1.3	3.4	1.1	1.8	0.3	2.4
Indegree	11.9	1.0	0.8	0.5	0.8	2.6	0.0	1.3
Type	Receiver	Ordinary	Ordinary	Ordinary	Ordinary	Ordinary	Transmitter	Ordinary
Degree Centrality	11.9	1.3	2.0	3.9	1.9	4.4	0.3	3.6
Betweenness Centrality	441.4225108225107	0.4	13.573809523809523	23.814285714285706	20.933766233766224	28.926190476190474	0	35.479437229437224
Closeness Centrality	24	12.5	16.166666666666664	15.833333333333334	15.166666666666668	16.833333333333336	11.083333333333334	16.5

(Graph) Graph Indices

Complexity	0.2000
Density	0.0996
Hierarchy Index	0.0220

Figure 7. Screenshot of the "Structural Analysis" tab.

Table 3. Centrality indices (Degree and Betweenness) for all concepts, calculated in FCMWizard.

Concept	Centrality			Centrality			Centrality	
	Degree	Betweenness		Degree	Betweenness		Degree	Betweenness
C1	11.9	441.4	C11	6.3	63.6	C21	1.8	11.3
C2	1.3	0.4	C12	3.9	30.8	C22	1.5	2.6
C3	2.0	13.5	C13	0.6	0	C23	0.3	0
C4	3.9	23.8	C14	5.1	125.0	C24	4.3	15.3
C5	1.9	20.9	C15	0.5	0	C25	5.5	49.7
C6	4.4	28.9	C16	4.5	106.9	C26	1.0	7.3
C7	0.3	0	C17	1.8	5.6	C27	1.3	1.5
C8	3.6	35.4	C18	5.8	45.5	C28	1.9	0.4
C9	3.8	3.8	C19	3.8	21.5	C29	3.0	1.5
C10	2.0	0	C20	5.0	29.6			

4.3. Scenario Analysis

FCMWizard also provides users with the ability to perform "what-if" scenario analysis after they specified the desired parameters of the FCM model to simulate [51,52]. The initial stimulus state vector, the inference rule's type, the transfer function, its learning parameter, and the number of iterations or the convergence step need to be properly defined. Users are also offered the open/closed lock option to set and keep the value of a concept unchained throughout iterations, as being "clamped" [25].

An example of a simulation model run by the tool for the examined case is presented below and regards the investigation of the impact that concept C11-"Government incentives" has on other concepts as well as on the examined system, overall.

Before the simulation process takes place, a baseline scenario needs to be conducted, so in the next step there will be a comparison between the results of the three scenarios and those resulting from the state of the model where all the initial values of concepts are zero (baseline scenario). In Figure 8, an exemplary table with the first 20 concepts values in all iterations steps, regarding the baseline scenario, is presented, while Figure 9 depicts the corresponding graph showing the steady state of every concept comprising this FCM model.

C1	C2	C3	C4	C5	C6	C7	C8	C9	C10	C11	C12	C13	C14	C15	C16	C17	C18	C19	C20
0.5	0.5	0.5	0.5	0.5	0.5	0.5	0.5	0.5	0.5	0.5	0.5	0.5	0.5	0.5	0.5	0.5	0.5	0.5	0.5
0.9925	0.7311	0.7058	0.6792	0.7058	0.4838	0.6225	0.7549	0.6792	0.7311	0.7191	0.7667	0.5927	0.8442	0.6225	0.5659	0.5622	0.4072	0.6614	0.9525
0.9995	0.8123	0.7891	0.7074	0.7712	0.3953	0.6506	0.8393	0.742	0.8285	0.7899	0.8563	0.6109	0.9209	0.6508	0.5366	0.5521	0.306	0.7232	0.98
0.9997	0.8351	0.8121	0.7027	0.7861	0.3535	0.6572	0.8618	0.7616	0.8519	0.8068	0.8788	0.617	0.9343	0.6572	0.5143	0.5415	0.2627	0.7481	0.9854
0.9998	0.8407	0.8174	0.6972	0.788	0.339	0.6586	0.8676	0.7672	0.8565	0.8102	0.8839	0.6197	0.937	0.6586	0.5057	0.537	0.249	0.755	0.9865
0.9998	0.842	0.8186	0.6946	0.7877	0.3346	0.6589	0.8691	0.7687	0.8574	0.8108	0.8851	0.6209	0.9376	0.6589	0.5029	0.5357	0.2452	0.7567	0.9867
0.9998	0.8423	0.8189	0.6936	0.7873	0.3334	0.659	0.8695	0.7691	0.8575	0.8109	0.8853	0.6213	0.9377	0.659	0.5021	0.5354	0.2443	0.7571	0.9867
0.9998	0.8424	0.8189	0.6933	0.7871	0.333	0.659	0.8696	0.7692	0.8576	0.8109	0.8854	0.6215	0.9377	0.659	0.5018	0.5354	0.2441	0.7572	0.9867
0.9998	0.8424	0.8189	0.6932	0.787	0.3329	0.659	0.8696	0.7692	0.8576	0.8109	0.8854	0.6215	0.9377	0.659	0.5017	0.5354	0.244	0.7573	0.9867
0.9998	0.8424	0.8189	0.6932	0.787	0.3329	0.659	0.8696	0.7692	0.8576	0.8109	0.8854	0.6215	0.9377	0.659	0.5017	0.5354	0.244	0.7573	0.9867

Figure 8. Screenshot of the FCM iterations steps in the baseline scenario from FCMWizard.

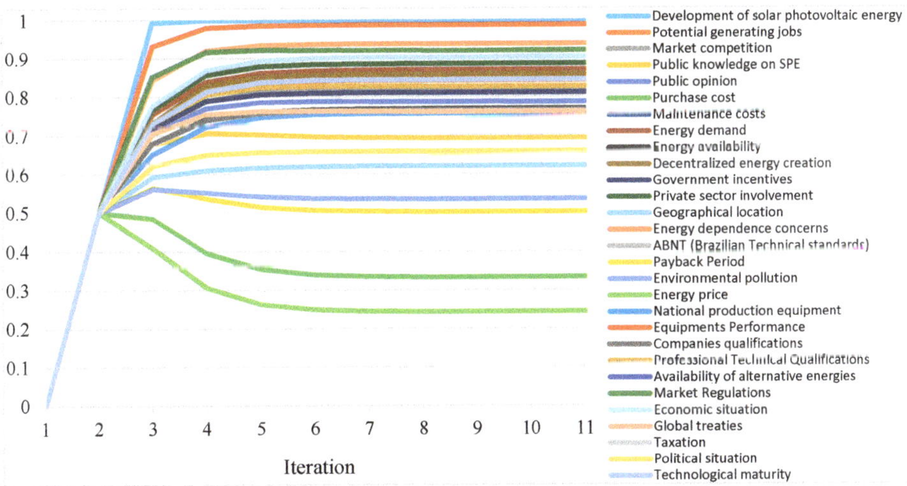

Figure 9. Screenshot of the graph regarding concepts' values in the Baseline scenario.

Let's now consider that concept "Government Incentives" (C11) is activated for the considered PSE problem, and the process of how this concept affects the other concepts needs to be investigated. For performing the scenario analysis, authors need to define a number of parameters like the inference rule, the transformation function, Lambda parameter, and either the number of iterations or the convergence step. All these simulation options and learning parameters are offered by the tool. In particular, it supports three FCM inference rules, four threshold functions, besides the possibility of customizing other parameters too [8].

In Figure 10, a screenshot of the options provided by "FCMWizard" is illustrated, regarding the parameters' configuration for the simulation process. Specifically, the initial value of the concept C11 was clamped to 1, the Modified Kosko's activation rule and the Sigmoid transformation function were selected, as well as the Lambda parameter and the convergence step were set to 1 and 0.0001 respectively.

The produced values for every concept of this scenario analysis example in the form of table and graph are depicted in the following figures (see Table 4 and Figure 11). The rest of the results are presented in Section 5.2, where the variety of functionalities and capabilities of the FCMWizard tool is further revealed.

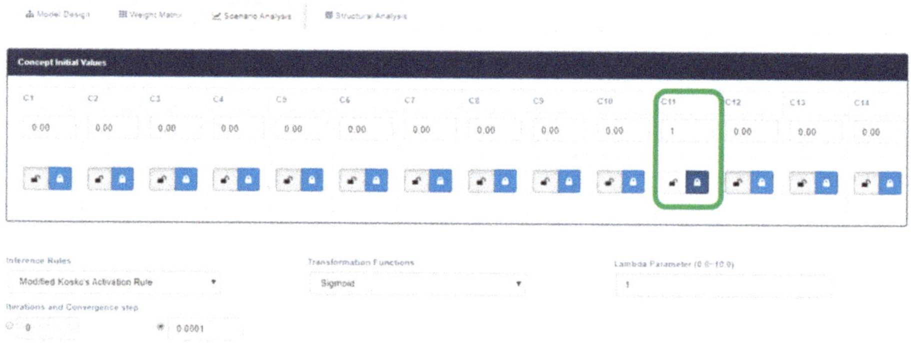

Figure 10. Screenshot of the Scenario analysis mode by FCM Wizard.

Table 4. Table with concepts' values for each iteration, produced from FCMwizard.

Title	C1	C2	C3	C4	C5	C6	C7	C8	C9	C10	C11
Iteration 0	0.661	0.679	0.5	0.5	0.5	0.347	0.5	0.5	0.5	0.5	1
Iteration 1	0.996	0.825	0.713	0.672	0.705	0.341	0.622	0.759	0.679	0.731	1
Iteration 2	0.999	0.854	0.793	0.700	0.770	0.306	0.650	0.840	0.742	0.828	1
Iteration 3	0.999	0.861	0.813	0.698	0.785	0.297	0.657	0.862	0.761	0.851	1
Iteration 4	0.999	0.862	0.818	0.694	0.787	0.296	0.658	0.867	0.767	0.856	1
Iteration 5	0.999	0.862	0.819	0.692	0.787	0.296	0.658	0.869	0.768	0.857	1
Iteration 6	0.999	0.862	0.819	0.69	0.787	0.296	0.659	0.869	0.769	0.857	1
Iteration 7	0.999	0.862	0.819	0.691	0.786	0.296	0.659	0.869	0.769	0.857	1
Iteration 8	0.999	0.862	0.819	0.691	0.786	0.296	0.659	0.869	0.769	0.857	1
Iteration 9	0.999	0.862	0.819	0.691	0.786	0.296	0.659	0.869	0.769	0.857	1

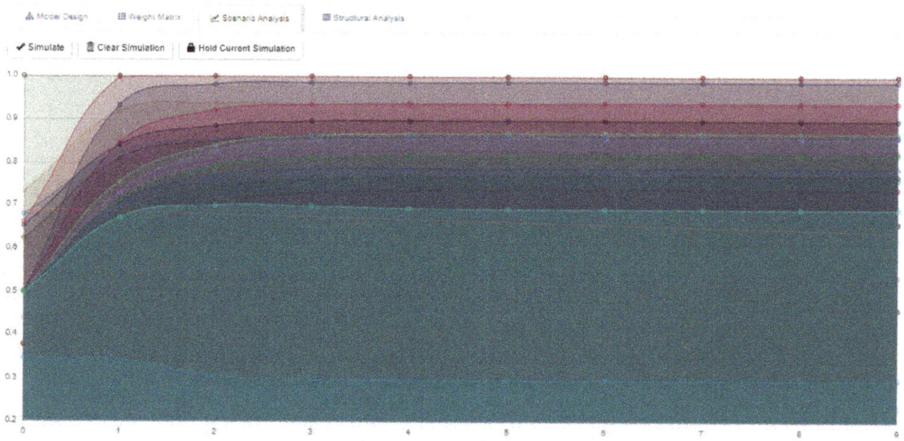

Figure 11. Graphical plot of convergence performed by FCMWizard.

Through this process, FCMWizard is proven to be a useful tool for designing FCM models and building simulations for different scenarios. It provides easy-to-use but powerful capabilities in modelling, analyzing and simulating complex systems with high uncertainty, and thus, it can be applied by policymakers for strategic decisions in various scientific domains. It should be noted that this web-based software is freely available to users upon a password request and is under continuous development.

5. Applying FCMWizard for Scenario Analysis

For a deeper understanding of the examined problem, such as the concept's behavior, their relations, and the extent to which one concept has an impact on others, an FCM simulation process is needed. This study employs the FCMWizard tool to simulate and analyze the problem [25]. This process entails the multiplication of the input vector with the adjacency matrix, and in the output vector a squashing function (Equation (2)) is applied each time as a threshold function, until the system reaches a stable state.

The process of simulation is performed by "clamping" the initial value of certain key concepts each time. This outcome is compared against a baseline scenario where the system reaches the steady state. After analyzing the deviations of concepts values between the baseline steady state and the outcome of the simulation process, researchers can interpret the impact of the key concepts on the system on a quantitative basis.

5.1. Scenario Development

The first established approach in scenario planning was the selection of the most important concepts (called decision concepts). Among the concepts selected to construct the studied FCM model (see Table 2), the researchers identified five concepts that could well affect the behavior of the system. These concepts (C4, C5, C11, C17, and C18), given by the programme participants and implementers, were selected as they were among the concepts with the highest centrality, having both in/out-degree values, while they are strongly connected with the objective of this scenario planning, which is the development of PSE (concept C1).

For the purposes of this study, three scenarios were developed through plausible combinations of the aforementioned decision concepts. To check the behavior of the system as a whole, we tested a set of combinations in which the concepts' values were activated and clamped to one. Thus, we create three scenarios: "Instability Scenario", "Disastrous" and "Government Incentives".

The selected scenarios with their concepts are briefly presented in the following table (Table 5).

Table 5. The key concepts of each scenario.

Scenarios	Concepts	
Scenario 1 (S1)	C18: Energy Price	
Scenario 2 (S2)	C4: Public Knowledge on PSE C5: Public opinion	C11: Government Incentives C17: Environmental pollution
Scenario 3 (S3)	C11: Government Incentives	

- Scenario 1 (S1) examined the effects of energy price (C18) in monetary terms.
- Scenario 2 (S2) presented the effects of public knowledge on SPE (C4) in terms of general awareness of the role and significance of PSE, public opinion (C5) considering the awareness of the society about the importance of environmental protection, government incentives (C11) in terms of taxation and financial grants, along with environmental pollution (C17) in terms of energy substances affecting the environment.
- Scenario 3 (S3) highlighted the effects of government incentives (C11) in terms of taxation and other financial benefits, granted by the government.

The first scenario, namely "Instability Scenario", reflects the recent history of Brazil, in which the country is undergoing economic and political instability. The current economic crisis in Brazil began in mid-2014, and this economic crisis was accompanied and intensified by a political crisis [53]. As a result of these crises, Dilma Rousseff, president of the time, suffered impeachment in 2016 [54]. The crisis also generated unemployment, which reached its peak in 2017 [55]. In response, the price of energy has risen dramatically. Consecutively, the concept "Energy price" was chosen to be the only concept examined in this scenario and so, its value is clamped to 1.

The "Disastrous" scenario reflects the shortage of electricity in Brazil, the blackouts, as well as some other environmental disasters, namely Mariana and Brumadinho [56]. Brazil has had to activate thermoelectric plants regularly, but this backup capacity was insufficient, and rationing had to be imposed. To avoid another "blackout crisis," the government decided to create a set of initiatives to encourage the growth of PSE, through the establishment of subsidized credit lines and tax exemptions. In 2015 and 2019, the mining dams named Brumadinho and Mariana, collapsed. Part of the demand for energy migrated to the PSE and the price of electricity provided by the government decreased. This scenario examines the impact the concepts "Public knowledge on PSE" and "Government incentives" have on the system dynamics and were accordingly activated and clamped to 1.

In the third, "Government Incentives" scenario, Government offers financial assistance to private businesses making investments through the use of economic incentives. Incentives can include tax abatements, tax revenue sharing, grants, infrastructure assistance, no or low-interest financing, free land, tax credits, and other financial resources. Thus, there seems to be more private sector involvement leading to a greater market competition, resulting in an overall energy price decrease. This could make the public feel more comfortable to consume more energy. In this scenario, the concept "Government Incentives" was clamped to 1, so the researchers can estimate through the simulation process the degree this concept affects the examined energy system.

5.2. Scenario Analysis for PV Solar Energy Sector

As it was reported in the "Scenario Analysis" section, the simulation process started after the baseline scenario was performed. Then, scenario analysis performed simulations for the selected three scenarios (Table 5). In this context, scenario 1 (S1) was devoted to increasing the decision concept C18—"Energy price" by "clamping" it to one. Scenario 2 (S2) studied the effects of the decision concepts C4—"Public Knowledge on PSE", C5—"Public opinion", C11—"Government Incentives" and C17—"Environmental pollution" by clamping the values of these concepts to one. In the end, Scenario 3 (S3) investigated how the increase of decision concept C11—"Government Incentives" affects the examined system, when "clamping" its values to one. The results for scenario analysis are illustrated in the following figures. The results are interpreted by comparing the relative change between this baseline scenario and the new steady state.

Figures 12–14 illustrate separately the outcomes of the three scenarios performed (S1, S2 and S3 respectively), with respect to the percentage of change (deviations from the steady state) for each concept. Thus, the authors are able to analyze and further interpret the results produced for the examined renewable energy system scenario, that will help policy makers select the right strategies in this direction. Negative values mean that the concepts have negative acceleration, whereas positive values imply that the concepts have positive acceleration. Worth mentioning is that negative change in the value of development of solar energy (C1) is regarded as a deceleration in deployment and diffusion of solar energy compared to the current state.

The end goal of the examined case study was set to be the development of the solar photovoltaic energy sector in Brazil, which is represented by the concept C1. Therefore, the scenario analysis mainly focuses on the impact that the decision concepts have on this output concept C1—"Development of solar photovoltaic energy", which is considered by the researchers as the main objective of this project. Furthermore, there is an extended discussion about how the rest of the decision concepts involved in scenario analysis, directly affect in a positive or negative way other significant concepts of the examined PSE sector.

5.3. Discussion of Scenario Analysis Results

The problem of Brazilian PSE sector development is tackled with the technique of FCMs using FCMWizard as a powerful simulation environment for policy making in the renewable energy domain. The researchers have the ability to form certain scenarios with the help of energy specialists, and stakeholders form the specific domain and conduct simulations in order to observe how the examined system responds to changes of certain concepts' values. Casting a thorough look on the results presented in the figures above (Figures 12–14), certain conclusions can emerge for each scenario conducted.

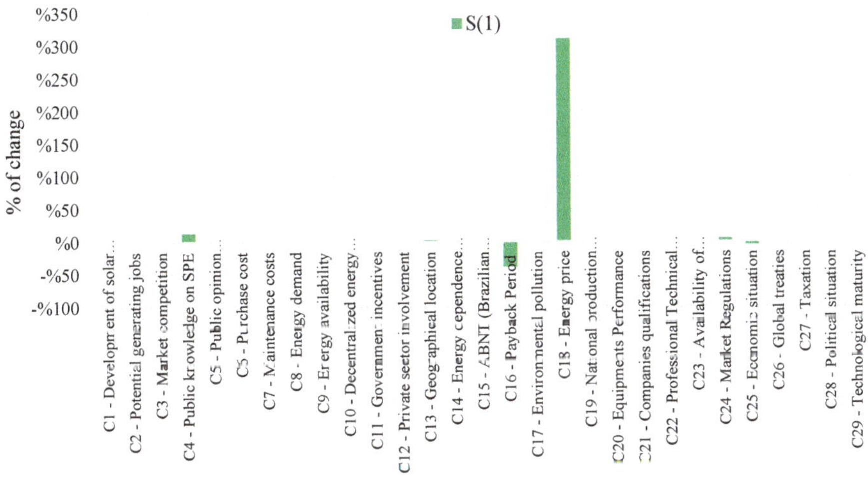

Figure 12. Percentage of change (deviation) for all key concepts when concept C18 is activated (S1).

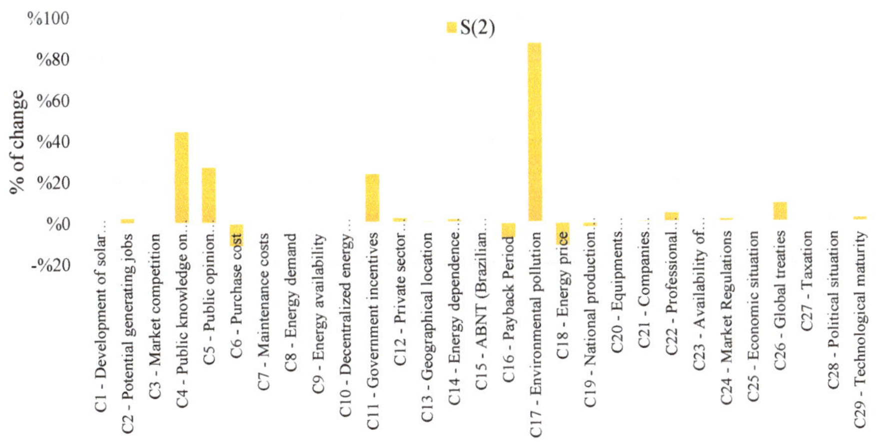

Figure 13. Percentage of change (deviation) for all key concepts when concepts C4, C5, C11, and C17 are activated (S2).

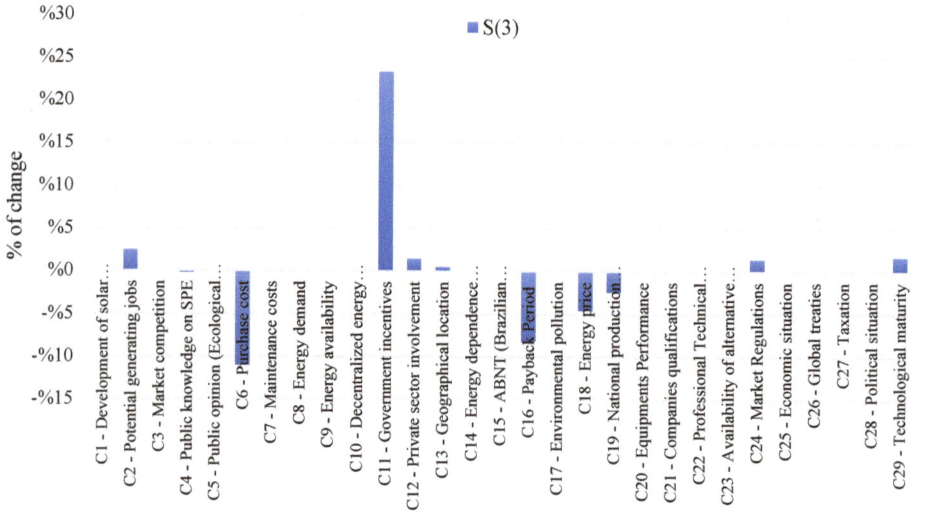

Figure 14. Percentage of change (deviation) for all key concepts when concept C11 is activated (S3).

The overall observations that were drawn from the figures above, focus on the following points:

- Considering the first conducted scenario (S1) and the "clamped" decision concept C18—"Energy price", the final values of decision concepts C1, C4, C13, C16, C24, and C25 have changed (increased or decreased) compared to those of the Baseline scenario, for the same number of iterations. Thus, it can be concluded that concept C18—"Energy price" mainly affects positively decision concepts C1, C4, C13, and C24, whereas concepts C16 and C25 are negatively affected.
- Following the same direction, the results of scenario S2 (see Figure 13) reveal that an overall increase in decision concepts C4, C5, C11, and C17 leads to a slight increase in the values of most concepts, while concepts C1, C2, C12, C22, and C25 seem to be positively affected to a greater extent. On the other hand, there is a noteworthy decrease in the final values of concepts C6, C16, C18, and C19.
- In the case of the third scenario and the clamped concept C11—"Government Incentives", the increase of its value is directly connected to an increase of the values of key concepts C2, C12, C13, C24, and C29. On the contrary, concept C11 affects inverse concepts C6, C16, C18, and C19.
- The participant concepts of Scenarios S1 and S2 have a significant positive impact on decision concept C1, which is the main outcome of this study, whereas Scenario 3 does not present any impact on that concept.
- When concept C11—"Government Incentives" is combined with other concepts, as participant concepts of the same scenario (S2), then it contributes to the increase of the decision concept C1 and the overall change of the examined system's state. When it performs individually, as a single concept of Scenario S3, then no impact on concept C1-"Development on PSE" and the overall system status are noticed.

Focusing on the first scenario (S1), outcomes showed that an increase in energy price led people to seek new energy sources alternatives and thus, energy demand was decreased, while public awareness of the SPE was increased, given that the payback period decreased. Furthermore, since the companies might have identified a very turbulent and disadvantageous market to exploit, a reduction in the private sector involvement is observed. Without the participation of the private sector, the purchase cost increased.

The conclusions for the second scenario (S2) are briefly presented as follows. Since environmental pollution has sharpened people's ecological awareness and knowledge on PSE, the Brazilian government creates various initiatives and programs to foster the growth of this sector, reducing the payback time of the investment. Part of the demand for energy migrated to the PSE and the price of electricity provided by the government, decreased. The purchase cost decreased accordingly.

In response to the third scenario (S3), where government incentives are considered, economic incentives and other financial resources, given by the government, can possibly cause a significant increase in private sector involvement and greater competition in the PSE market, resulting in a bigger employment capacity. A decrease in energy price is possible and consequently, the country's own inhabitants will feel more confident to consume more energy in their daily life.

Taking into consideration the key concepts of the three conducted scenarios, it emerges that government incentives are mostly able to change the stability of Brazil's Renewable Energy system. After a critical overview of the overall results, it is unlikely the PSE sector develops much, unless there are noteworthy political and economic changes in the status of the country. Moreover, the FCM-based simulation process has shown that Brazil's PSE development is dependent on economic parameters such as the incentives given by the government and is influenced by uncertain factors like public ecological awareness. In addition, the penetration of the private sector in the PSE can affect the dynamics of the system as well as have an impact on the everyday life of Brazilian citizens.

The conducted scenarios highlighted the system's uncertainty and showed to the policy makers how they can develop effective strategies for creating a more robust system that is free from possible future instabilities. The tool settings that are offered, can help users to conduct further scenarios and try different configurations by activating different variables, individually or in a set, as well as by changing the weighted interconnections among them, in order to see how the model reacts in different circumstances.

6. Conclusions

This study deals with the contribution of the FCM method with the help of an efficient simulation tool called FCMWizard, which has been developed to implement/deploy certain tasks such as modeling experts' knowledge, learning and simulating FCM-based complex systems, in the participatory domain. The main contribution of this work lies in the fact that the authors provide experts and stakeholders that belong to various scientific areas, a free under request, user-friendly and flexible, as regards the number of variables and weights, tool, which can be used anytime and in any discipline or domain, without time or memory restrictions. This software that uses certain learning algorithms, offers to users the ability to easily and intuitively design a model for a particular problem, as well as perform scenario planning for making strategic decisions and policy making in a simple, user-friendly environment, without any computational cost.

The innovation in this work is based on the fact that, decision-makers can apply the Fuzzy Cognitive Map method along with the proposed software tool to model experts' or stakeholders' perceptions and further conduct different scenarios implemented by the FCMWizard tool. In this context, policy-making can be performed in various domains and with different configurations of the tool, regarding the type of inference rules, the number of variables, and the weights values or concepts' selection in the scenario planning, thus, highlighting the generic applicability and usefulness of the new tool.

To show the functionalities and examine the performance of this flexible and interactive web-based software, the authors explored a case study concerning the development of Brazil's Photovoltaic Solar energy sector and a relevant scenario analysis was performed. The main outcomes of this research will help policymakers that belong to the Renewable Energy domain, determine certain strategies for better Supply Chain Management, contributing to optimal exploitation of the PSE sector and Brazil's overall socio-economic growth. In this context, a variety of simulation options and different learning parameters were used for proper configuration of the examined system, in terms of modeling, analyzing and simulating the FCM model. In particular, three FCM inference rules and four threshold functions can be selected by any type of user (experts, stakeholders, policy makers or others), while a number of other useful parameters are offered for customizing simulations.

Certain limitations of this work need to be considered in this section. For example, the provided tool is under continuous development to further include more and advanced functionalities, such as the straightforward use of fuzzy sets for FCM design, interval fuzzy values for weights, other efficient and state-of-the-art fuzzy methods (intuitionistic, type 2-fuzzy sets) for FCM model design, as well as other constraints on the concepts that constitute the aggregated FCM model.

Future work is directed towards further investigation of more scenarios and for different scientific domains. Moreover, for validation purposes, this method could be applied in a respective Renewable Energy domain, where more variables will be considered, while the producing results would be further analyzed and compared to those of the current study. Also, future work is devoted to the development of the data-based construction of FCMs where no qualitative or expert/stakeholder knowledge is available to design an FCM model for scenario analysis.

Author Contributions: Conceptualization, K.P.; methodology, K.P. and G.C.; software, K.P.; validation, K.P., G.C., E.I.P. and D.B.; formal analysis, K.P. and G.C.; investigation, K.P. and E.I.P.; resources, G.C.; data curation, G.C. and K.P.; writing—original draft preparation, K.P.; writing—review and editing, E.I.P. and D.B.; visualization, K.P.; supervision, E.I.P., and G.S.; project administration, E.I.P. and D.B. All authors have read and agreed to the published version of the manuscript.

Funding: This research received no external funding.

Conflicts of Interest: The authors declare no conflict of interest.

References

1. Kosko, B. Fuzzy cognitive maps. *Int. J. Man Mach. Stud.* **1986**, *24*, 65–75. [CrossRef]
2. Papageorgiou, E.I. Learning Algorithms for Fuzzy Cognitive Maps: A Review Study. *IEEE Trans. Syst. Man Cybern. Part C* **2012**, *42*, 150–163. [CrossRef]
3. Reckien, D.; Wildenberg, M.; Bachhofer, M. Subjective Realities of Climate Change: How Mental Maps of Impacts Deliver Socially Sensible Adaptation Options. *Sustain. Sci.* **2013**, *8*, 159–172. [CrossRef]
4. Groumpos, P.P. Fuzzy Cognitive Maps: Basic Theories and Their application to Complex Systems. *Comput. Sci.* **2010**. [CrossRef]
5. Mateou, N.H.; Andreou, A.S. A framework for developing intelligent decision support systems using evolutionary fuzzy cognitive maps. *J. Intell. Fuzzy Syst.* **2008**, *19*, 151–170.
6. Papageorgiou, E.I. A new methodology for decisions in medical informatics using fuzzy cognitive maps based on fuzzy rule-extraction techniques. *Appl. Soft Comput.* **2011**, *11*, 500–513. [CrossRef]
7. Salmeron, J.L. Supporting decision makers with fuzzy cognitive maps. *Res. Technol. Manag.* **2009**, *52*, 53–59. [CrossRef]
8. Papageorgiou, E.I.; Papageorgiou, K.; Dikopoulou, Z.; Mouhrir, A. A web-based tool for Fuzzy Cognitive Map Modeling. In Proceedings of the 9th International Congress on Environmental Modelling and Software (iEMSs), Fort Collins, CO, USA, 24–28 June 2018.
9. Amer, M.; Jetter, A.; Daim, T. Development of fuzzy cognitive map (FCMs)-based scenarios for wind energy. *Int. J. Energy Sect. Manag.* **2011**, *5*, 564–584. [CrossRef]
10. Kok, K. The potential of Fuzzy Cognitive Maps for semi-quantitative scenario development, with an example from Brazil. *Glob. Environ. Chang.* **2009**, *19*, 122–133. [CrossRef]
11. Van Vliet, M.; Kok, K.; Veldkamp, T. Linking stakeholders and modelers in scenario studies: the use of Fuzzy Cognitive Maps as a communication and learning tool. *Futures* **2010**, *42*, 1–14. [CrossRef]
12. Jetter, A.; Schweinfort, W. Building scenarios with Fuzzy Cognitive Maps: An exploratory study of solar energy. *Futures* **2011**, *43*, 52–66. [CrossRef]
13. Karagiannis, I.E.; Groumpos, P.P. Modeling and analysis of a hybrid-energy system using fuzzy cognitive maps. In Proceedings of the 21st Mediterranean Conference on Control and Automation, Chania, Greece, 25–28 June 2013; pp. 257–264.

14. Huang, S.; Lo, S.; Lin, Y. Application of a fuzzy cognitive map based on a structural equation model for the identification of limitations to the development of wind power. *Energy Policy* **2013**, *63*, 851–861. [CrossRef]
15. Ziv, G.; Watson, E.; Young, D.; Howard, D.C.; Larcom, S.T.; Tanentzap, A.J. The potential impact of Brexit on the energy, water and food nexus in the UK: a fuzzy cognitive mapping approach. *Appl. Energy* **2018**, *210*, 487–498. [CrossRef]
16. Schoemaker, P.J. Scenario planning: a tool for strategic thinking. *Sloan Manag. Rev.* **1995**, *36*, 25–50.
17. Stylios, C.; Groumpos, P.P. The challenge of modelling supervisory systems using fuzzy cognitive maps. *J. Intell. Manuf.* **1998**, *9*, 339–345. [CrossRef]
18. Azadeh, A.; Arab, R.; Behfard, S. An adaptive intelligent algorithm for forecasting long term gasoline demand estimation: The cases of USA, Canada, Japan, Kuwait and Iran. *Expert Syst. Appl.* **2010**, *37*, 7427–7437. [CrossRef]
19. Azadeh, A.; Saberi, M.; Gitiforouz, A.; Saberi, Z. A hybrid simulation adaptive-network-based fuzzy inference system for improvement of electricity consumption estimation. *Expert Syst. Appl.* **2009**, *36*, 11108–11117. [CrossRef]
20. Borrie, D.; Ozveren, C.S. The electric power market in the United Kingdom: Simulation with adaptive intelligent agents and the use of fuzzy cognitive maps as an inference engine. In Proceedings of the 39th International Universities Power Engineering Conference, 2004. UPEC 2004, Bristol, UK, 6–8 September 2004; Volume 3, pp. 1150–1154.
21. Hol, M.; Bilgin, A. Design of a Fuzzy Logic Based Framework for Comprehensive Anomaly Detection in Real-World Energy Consumption Data. In *Benelux Conference on Artificial Intelligence*; Springer: Cham, Switzerland, 2017; pp. 121–136.
22. Antosiewicz, M.; Nikas, A.; Szpor, A.; Witajewski-Baltvilks, J.; Doukas, H. Pathways for the transition of the Polish power sector and associated risks. *Environ. Innov. Soc. Transit.* **2019**. [CrossRef]
23. Nikas, A.; Ntanos, E.; Doukas, H. A semi-quantitative modelling application for assessing energy efficiency strategies. *Appl. Soft Comput.* **2019**, *76*, 140–155. [CrossRef]
24. Nikas, A.; Doukas, H.; van der Gaast, W.; Szendrei, K. Expert views on low-carbon transition strategies for the Dutch solar sector: A delay-based fuzzy cognitive mapping approach. *IFAC PapersOnLine* **2018**, *51*, 715–720. [CrossRef]
25. Alipour, M.; Hafezi, R.; Papageorgiou, E.I.; Hafezi, M.; Alipour, M. Characteristics and scenarios of solar energy development in Iran: Fuzzy cognitive map-based approach. *Renew. Sustain. Energy Rev.* **2019**, *116*. [CrossRef]
26. Gray, S.A.; Gray, S.; Cox, L.J.; Henly-Shepard, S. Mental modeler: a fuzzy-logic cognitive mapping modeling tool for adaptive environmental management. In Proceedings of the 2013 46th Hawaii International Conference on System Sciences, Wailea, HI, USA, 7–10 January 2013; pp. 965–973.
27. Bachhofer, M.; Wildenberg, M. FCMappers—Disconnecting the Missing Link. Available online: http://www.fcmappers.net/joomla/ (accessed on 10 February 2020).
28. Napoles, G.; Leon, M.; Grau, I.; Vanhoof, K. Fuzzy cognitive maps tool for scenario analysis and pattern classification. In Proceedings of the 2017 IEEE 29th International Conference on Tools with Artificial Intelligence (ICTAI), Boston, MA, USA, 6–8 November 2017; pp. 644–651.
29. Margaritis, M.; Stylios, C.; Groumpos, P.P. Fuzzy cognitive map software. In Proceedings of the 10th International Conference on Software, Telecommunications and Computer Networks SoftCom, Split, Dubrovnik, Croatia, Ancona, Venice, Italy, 8–11 October 2002; pp. 8–11.
30. De Franciscis, D. JFCM: a java library for FuzzyCognitive maps. In *Fuzzy Cognitive Maps for Applied Sciences and Engineering*; Springer: Berlin/Heidelberg, Germany, 2014; pp. 199–220.
31. Glykas, M. *Fuzzy Cognitive Maps: Advances in Theory, Methodologies, Tools and Applications*; Springer: Berlin/Heidelberg, Germany, 2010.
32. Papageorgiou, K.; Singh, P.K.; Papageorgiou, E.; Chudasama, H.; Bochtis, D.; Stamoulis, G. Fuzzy Cognitive Map-Based Sustainable Socio-Economic Development Planning for Rural Communities. *Sustainability* **2019**, *12*, 305. [CrossRef]
33. Singh, P.K.; Papageorgiou, K.; Chudasama, H.; Papageorgiou, E.I. Evaluating the Effectiveness of Climate Change Adaptations in the World's Largest Mangrove Ecosystem. *Sustainability* **2019**, *11*, 6655. [CrossRef]

34. Trappey, A.J.C.; Trappey, C.V.; Hao, T.; Liu, P.H.Y.; Shin-Je, L.; Lin, L. The determinants of photovoltaic system costs: an evaluation using a hierarchical learning curve model. *J. Clean. Prod.* **2016**, *112*, 1709–1716. [CrossRef]
35. Ferreira, A.; Kunh, S.S.; Fagnani, K.C.; De Souza, T.A.; Tonezer, C.; Dos Santos, G.R.; Coimbra-Araújo, C.H. Economic overview of the use and production of photovoltaic solar energy in brazil. *Renew. Sustain. Energy Rev.* **2018**, *81*, 181–191. [CrossRef]
36. Papageorgiou, E.I.; Salmeron, J.L. Methods and Algorithms for Fuzzy Cognitive Map-based Modeling. *Fuzzy Cogn. Maps Appl. Sci. Eng.* **2014**.
37. Papageorgiou, E.I.; Subramanian, J.; Karmegam, A.; Papandrianos, N. A risk management model for familial breast cancer: A new application using Fuzzy Cognitive Map method. *Comput. Methods Programs Biomed.* **2015**, *122*, 123–135. [CrossRef]
38. Nikas, A.; Doukas, H. Developing robust climate policies: a fuzzy cognitive map approach. In *Robustness Analysis in Decision Aiding, Optimization, and Analytics*; Springer: Berlin/Heidelberg, Germany, 2016; pp. 239–263.
39. Rocha, B.C.; Ferreira, P.P.F.V.; Dias, D.H.N.; Borba, B.S.M.C. Economic evaluation of photovoltaic microgenaration in the Southeast region of Brazil. In Proceedings of the 2018 Simposio Brasileiro de Sistemas Eletricos (SBSE), Niteroi, Brazil, 12–16 May 2018; pp. 1–6.
40. Lenz, A.M.; Colle, G.; de Souza, S.M.N.; Prior, M.; Nogueira, C.E.C.; Santos, R.F.; Friedrish, L.; Secco, D. Evaluation of three systems of solar thermal panel using low cost material, tested in Brazil. *J. Clean. Prod.* **2017**, *167*, 201–207. [CrossRef]
41. Martins, F.R.; Pereira, E.B.; Abreu, S.L. Satellite-derived solar resource maps for Brazil under SWERA project. *Solar Energy* **2007**, *81*, 517–528. [CrossRef]
42. Warren, B. *Renewable Energy Country Attractiveness Index (RECAI)*; Ernst Young: London, UK, 2015.
43. Ministério de Minas e Energia (MME) Energia Solar no Brasil e no Mundo. Available online: http://gesel.ie.ufrj.br/app/webroot/files/IFES/BV/mme68.pdf (accessed on 2 October 2019).
44. Wang, Z.; Wei, W. External cost of photovoltaic oriented silicon production: a case in China. *Energy Policy* **2017**, *107*, 437–447. [CrossRef]
45. Zilli, B.; Miguel Lenz, A.; Souza, S.; Secco, D.; Nogueira, C.; Ando Junior, O.H.; Nadaleti, W.; Siqueira, J.; Gurgacz, F. Performance and effect of water-cooling on a microgeneration system of photovoltaic solar energy in Paraná, Brazil. *J. Clean. Prod.* **2018**, *192*. [CrossRef]
46. Paulo, A.F.; Porto, G.S. Evolution of collaborative networks of solar energy applied technologies. *J. Clean. Prod.* **2018**, *204*, 310–320. [CrossRef]
47. Schoemaker, P.J. When and how to use scenario planning: a heuristic approach with illustration. *J. Forecast.* **1991**, *10*, 549–564. [CrossRef]
48. Brown, R.G. *Smoothing, Forecasting and Prediction of Discrete Time Series*; Prentice-Hall: Englewood Cliffs, NJ, USA, 1963.
49. Özesmi, U.; Özesmi, L.O. Ecological models based on people's knowledge: a multi-step fuzzy cognitive mapping approach. *Ecol. Model.* **2004**, *176*, 43–64. [CrossRef]
50. Gray, S.R.J.; Gagnon, A.; Gray, S.; O'Dwyer, B.; O'Mahony, C.; Muir, D.; Devoy, R.J.N.; Falaleeva, M.; Gault, J. Are coastal managers detecting the problem? Assessing stakeholder perception of climate vulnerability using Fuzzy Cognitive Mapping. *Ocean. Coast. Manag.* **2014**, *94*, 74–89. [CrossRef]
51. Kosko, B. *Neural Networks and Fuzzy Systems: A Dynamical Systems Approach to Machine Intelligence*; Prentice-Hall, Inc.: Englewood Cliffs, NJ, USA, 1992; ISBN 0-13-612334-1.
52. Papageorgiou, E.; Kontogianni, A. Using Fuzzy Cognitive Mapping in Environmental Decision Making and Management: A Methodological Primer and an Application. In *International Perspectives on Global Environmental Change*; IntechOpen: London, UK, 2012; ISBN 978-953-307-815-1.
53. Holland, M. Fiscal crisis in Brazil: causes and remedy. *Braz. J. Political Econ.* **2019**, *39*. [CrossRef]
54. Nunes, F.; Melo, C.R. Impeachment, Political Crisis and Democracy in Brazil. *Revista de Ciencia Política* **2017**, *37*, 281–304. [CrossRef]

55. *International Labour Organization Labour Overview 2017, Latin America and the Caribbean*; ILO Regional Office for Latin America and the Caribbean: Lima, Peru, 2017.
56. Torinelli, V.; Silva, A., Jr.; Souza, A.; Gomes, S.; Andrade, J. The impacts of environmental disasters on share value and sustainability index: an analysis of Vale company facing Mariana and Brumadinho disasters—Brazil. *Latin Am. J. Manag. Sustain. Dev.* **2020**, *5*, 75–94.

energies

MDPI

Article

A Cloud-Based In-Field Fleet Coordination System for Multiple Operations

Caicong Wu [1,2], Zhibo Chen [1], Dongxu Wang [1], Bingbing Song [1], Yajie Liang [1], Lili Yang [1,*] and Dionysis D. Bochtis [3,*]

1 College of Information and Electrical Engineering, China Agricultural University, Beijing 100083, China; wucc@cau.edu.cn (C.W.); chenzb@cau.edu.cn (Z.C.); Dongxu_Wang@cau.edu.cn (D.W.); songbing@cau.edu.cn (B.S.); liangyajie@cau.edu.cn (Y.L.)

2 Key Laboratory of Remote Sensing for Agri-Hazards, Ministry of Agriculture, Beijing 100083, China

3 Institute for Bio-economy and Agri-technology (IBO), Centre for Research and Technology—Hellas (CERTH), 6th km Charilaou-Thermi Rd, GR 57001 Thermi, Thessaloniki, Greece

* Correspondence: llyang@cau.edu.cn (L.Y.); d.bochtis@certh.gr (D.D.B.)

Received: 16 November 2019; Accepted: 5 February 2020; Published: 11 February 2020

Abstract: In large-scale arable farming, multiple sequential operations involving multiple machines must be carried out simultaneously due to restrictions of short time windows. However, the coordination and planning of multiple sequential operations is a nontrivial task for farmers, since each operation may have its own set of operational features, e.g., operating width and turning radius. Taking the two sequential operations—hoeing cultivation and seeding—as an example, the seeder has double the width of the hoeing cultivator, and the seeder must remain idle while waiting for the hoeing cultivator to finish two rows before it can commence its seeding operation. A flow-shop working mode can coordinate multiple machines in multiple operations within a field when different operations have different implement widths. To this end, an auto-steering-based collaborative operating system for fleet management (FMCOS) was developed to realize an in-field flow-shop working mode, which is often adopted by the scaled agricultural machinery cooperatives. This paper proposes the structure and composition of the FMCOS, the method of operating strip segmenting, and a new algorithm for strip state updating between successive field operations under an optimal strategy for waiting time conditioning between sequential operations. A simulation model was developed to verify the state-updating algorithm. Then, the prototype system of FMCOS was combined with auto-steering systems on tractors, and the collaborative operating system for the server was integrated. Three field experiments of one operation, two operations, and three operations were carried out to verify the functionality and performance of FMCOS. The results of the experiment showed that the FMCOS could coordinate in-field fleet operations while improving both the job quality and the efficiency of fleet management by adopting the flow-shop working mode.

Keywords: agricultural machinery; fleet management; auto-steering system; collaborative operating system; flow-shop; simulation; field experiment

1. Introduction

Agricultural fleet management is an important research domain that concerns resource allocation, operation scheduling, routing, communication, and real-time data acquisition [1]. Seamless communications and collaboration across spatial, temporal, operational, and machinery domains are a particular direction toward fleet management in scaled agricultural machinery service organization [2]. Sørensen and Bochtis [3] proposed that the total supply chain of large-scale harvesting has four levels of collaboration, i.e., in-field logistics, inter-field logistics, inter-sector logistics, and inter-regional logistics. In terms of machinery collaboration, Wang et al. [2] proposed that in-field logistics can be subdivided into three levels, i.e., multi-operation coordination, multi-machinery coordination, and multi-factor coordination [4]. Typical cases of multi-operation collaboration include the master–slave approach [5–7] and leader–follower approach [8–10], which illustrate multiple machines working in parallel for different but related operating tasks. These systems refer to identical operations. Other approaches have been proposed for sequential operations. As an example, in [11], a multi-operation collaboration system was proposed for six sequential operations in a potato production system—namely, bed forming, stone separation, planting, reloading unit, spraying, and harvesting. In parallel, algorithms for optimizing in-field route planning have been developed in various studies [12–15]. Furthermore, the flow-shop working mode of agricultural machinery operation is somewhat more complex [16–19] when multiple units of machinery and multiple operations exist within one field during the same period [20].

For the implementation of all the approaches mentioned above, Real-Time Kinematic–Global Navigation Satellite System (RTK-GNSS)-based automated steering systems can provide automatic guidance for single agricultural machines with one-inch repeatability [21,22]. Auto-steering technology can reduce overlaps and skips and can reduce the dependence upon skilled operators [23]. Therefore, fusion of the efficient flow-shop working mode and advanced auto-steering technologies can improve the field operation efficiency of agricultural machinery. To realize this fusion, the guidance lines followed by the auto-steering systems and the operating progress should be shared among the agricultural machinery units of the fleet in real time, so as to allow the operators to choose the appropriate operating strips for working. It is worth mentioning that different agricultural machinery operations have differing operating strips (in terms of number and size) and, therefore, different guidance lines, since different agricultural machines probably carry different implements with different operating widths. Most auto-steering systems can share guidance lines between different agricultural machinery units for the same operation through U disk copying, radio transfer, and/or server exchange. For instance, Cao et al. [24] built a Web-GIS (Geographic Information System) to realize collaboration among multiple agricultural machinery units, and the John Deere company utilizes an MTG (Modular Telematics Gateway) intelligent gateway and JDLink software to share guidance lines between different agricultural machinery units [25]. To support multi-operation collaboration, Wang et al. [2] designed a cloud-terminal-based coordinative operating system for in-field fleet management. The authors gave the definition of an operating strip to replace the path [26], made use of intervals for strip identification, and proposed an algorithm for strip state updating between adjacent machinery operations. The research demonstrated how to calculate the waiting time or flow step accurately to reduce the delay between adjacent machinery operations [2,19]. However, the validation of the system took place through simulation, while the strip state-updating algorithm provided a feasible—yet not optimized—procedure.

In order to optimize the state-updating algorithm and to validate the functionality of the work presented in Wang et al. [2] under real field experimental conditions, a new collaborative operating system for fleet management (FMCOS) was redesigned and developed, which can support all the operating stages of a field crop and incorporates an upgraded algorithmic process for strip state updating that simplifies the calculation process. The algorithm was tested through simulations using Matlab (MathWorks®) before the development of FMCOS. Finally, three field experiments were carried out to validate FMCOS and to test its functionality and performance.

2. Description of the Fleet Management System

2.1. System Composition

Typically, a complete cropping cycle consists of water management, land cultivation, seeding, nutrient management, pest management, and harvesting, as illustrated in Figure 1 [27]. For scaled agricultural machinery cooperatives, adoption of the flow-shop working mode in the phase of land cultivation is needed in order to improve the operating efficiency in a single field. Therefore, FMCOS should not only service the whole production cycle of the field crop, but also support the in-field flow-shop working mode for some specific operating phases. For this purpose, the auto-steering systems, a fixed GNSS base station, and a database that can manage the parameters of the field area and implements were deployed as a system. The baseline in each field can be used to generate guidance lines for all related machinery based on the width of the implement. FMCOS mainly includes two components: the GNSS-based auto-steering systems installed on the agricultural machinery and the cloud-based operation coordination system.

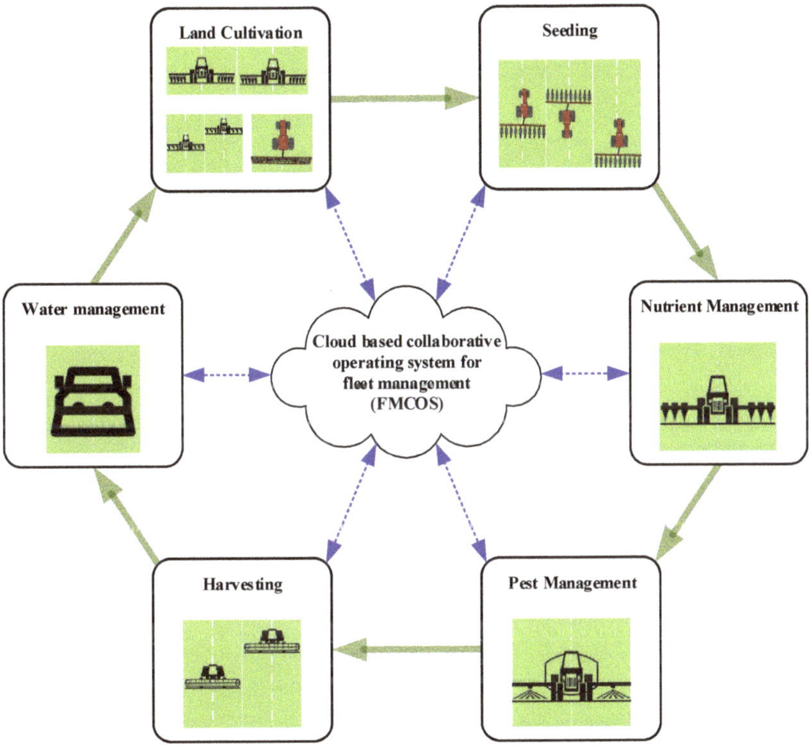

Figure 1. Collaborative application for the production cycle of field crop planting and management.

2.2. Operating Strip Segmentation

As shown in Figure 2, the field is segmented into strips for each machine. The width of the strip equals the width of the implement. Here, S_i^k refers to the ith strip of the kth operation, O^k. Strips are numbered from left to right. U_j^k are used to describe the machinery units.

A state shows the process progress in a strip by a machine and has four classes: *unready* (0), *ready* (1), *doing* (2), and *done* (3) (Table 1). When the task begins, the state of the strips that belong to the first operation is *ready*, and the machine arranged for the first operation can enter each of the strips for the first operation.

While the fleet is operating in the field, the strip states are updated in real time from the first operation to the last operation in turn. For a single operation, the strip states are updated from the left to the right. When the server gets the updated information of the operating strip from the Android application, the algorithm successively updates the strip state of the current operation according to the prior operating strip state.

Figure 2. Expressions of operation tasks, operating strips, and machinery units.

Table 1. Definitions of operating strip states.

State	Color	Value	Definition
unready	(grey)	0	The strip is not ready for operation.
ready	(olive)	1	The strip is ready for operation.
doing	(light green)	2	The strip is being processed by a machine.
done	(dark green)	3	The strip has been processed by a machine.

Figure 3 shows two operations, O^{k+1} and O^k. O^{k+1} is the operation after O^k, and the state of S_3^{k+1} of O^{k+1} is determined by three relevant strips of O^k, i.e., S_4^k, S_5^k, and S_6^k. This is because at the same location in the field, only when the related strips of O^k have been finished can the machinery for O^{k+1} enter the location. Therefore, the state (*ready* or *unready*) of O^{k+1} is determined by that of O^k. In other words, if, and only if, the state of the three strips becomes *done*, the state of S_3^{k+1} can be updated from *unready* to *ready*.

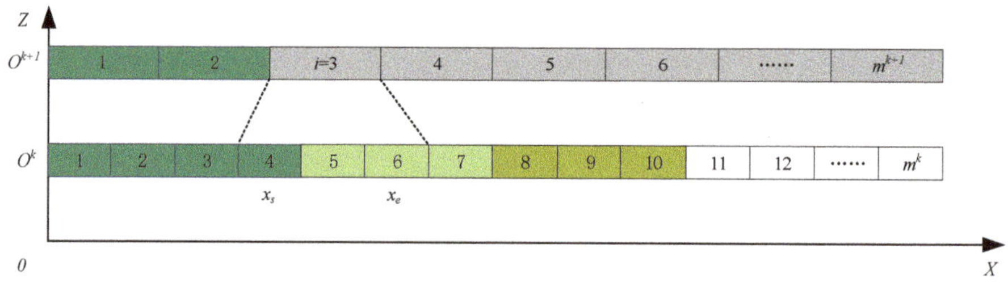

Figure 3. Operating strip state updating.

In general, to update the current strip state, the starting ID (x_s^k) and ending ID (x_e^k) of the prior operating strips should be determined. The two IDs and the total number (n) of related strips can be calculated by Formula (1). The aim of Formula (1) is to find all the strips of O^k that are related to O^{k+1}; this is equivalent to finding the IDs of the far-left strip and the far-right strip of O^k, which are related to O^{k+1}.

$$\begin{cases} x_s^k = int(\frac{w^{k+1}\times(i-1)}{w^k}) + 1 \\ x_e^k = int(\frac{w^{k+1}\times i}{w^k}) + 1 \\ n = x_e^k - x_s^k + 1 \end{cases} \tag{1}$$

With the above parameters, the average state value of the objective operating strip can be calculated by Formula (2), where $\overline{V}$ is the mean of the state values of all the related strips of O^k:

$$\overline{V} = \sum_{j=x_s^k}^{x_e^k} V_j^k \div n \tag{2}$$

Since 3 is the maximum value of the operating strip state, when $\overline{V} = 3$, that means that the state values of all the related strips are 3 (*done*). This is the condition required to change the strip state of the later operation (from *unready* to *ready*, or from 0 to 1), and the state of S_i^{k+1} can then be updated from *unready* to *ready*.

2.3. Sequential Operations Connection

2.3.1. Field Partitioning

For single-operation tasks, the field can be divided into partitions (sub-fields) according to the number of machinery units, and no optimization is needed within a partition. For multi-operation tasks, a partitioning strategy is also used for the flow-shop working mode. For instance, for the first operation task, all of the machines should work together from one side of the field to the other side, and each machine should occupy several operating strips as a temporary partition (Figure 4). When the machines finish a temporary partition, they move to the next temporary partition. Therefore, in each temporary partition, there will be no collisions between the multiple machinery units.

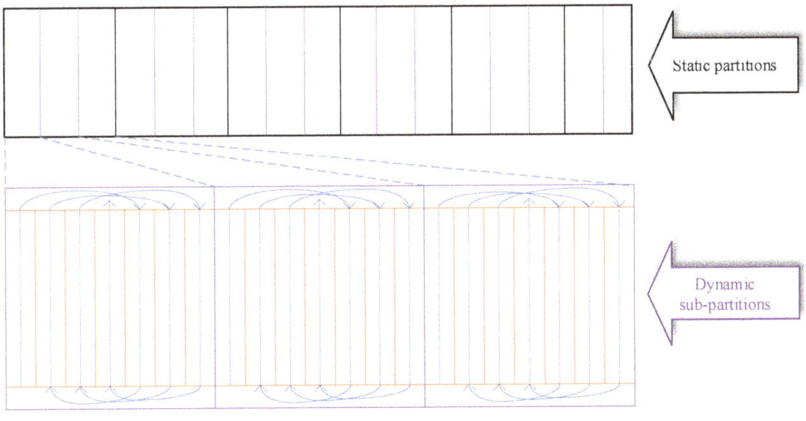

Figure 4. Dynamic partitioning for several sets of machinery units.

In practical work, the partitions for each operation are related to the number of machines, which is different in each operation. Different operations also have different work widths; thus, the partitioning for each operation is dynamic. Figure 4 shows an example of a fieldwork pattern with three sets of agriculture machinery. The entire field area is divided into six static partitions, and each partition is divided into three dynamic sub-partitions (excluding the last partition on the right). Each sub-partition is allocated to a single machine unit.

This study adopted the "first turn skip pattern" (FSP) [28] as the fieldwork pattern, which provides the highest savings in terms of in-field traversing distance among the most common pre-defined fieldwork patterns [29].

2.3.2. Operations Scheduling

The next step of the method includes the calculation of the optimal waiting time between two adjacent operations with a final goal to reduce the overall operating time. As shown in Figure 5, T^5 is the overall operating time of the fleet, and for its minimization, it is necessary to shorten the waiting times (T^k_{wait}) between any two adjacent operations.

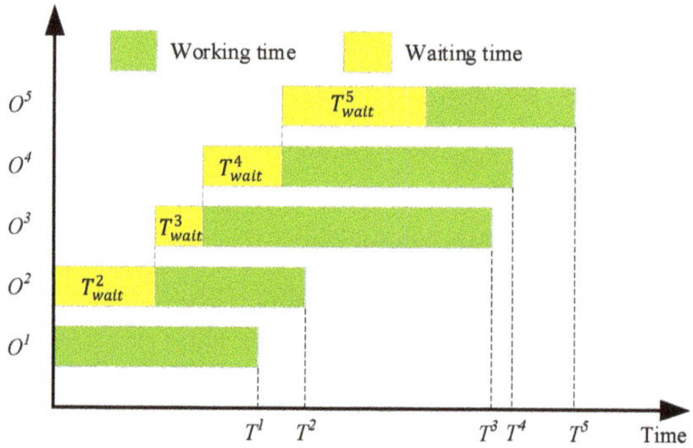

Figure 5. Waiting time and working time of each operation.

To calculate the waiting time, a new parameter, c^k, was defined to represent the operating capacity per unit time of the kth operation:

$$c^k = n^k \cdot v^k \cdot w^k \quad (3)$$

where n^k, v^k, and w^k are the number, velocity, and width, respectively, of the machinery unit of the kth operation.

Figure 6 shows the optimal waiting time between two adjacent operations. In Figure 6, each color represents an operation, and the horizontal line segment represents the duration within a single static partition. The left oblique line and the right oblique line go through the starting points and ending points of the horizontal line segments. To avoid collisions during operation, the left oblique line of the current operation should not intersect the right oblique line of the previous operation. Therefore, T^k_{wait} is determined by the relative relationship of the two slopes of the above two lines and can be calculated using Formula (4):

$$T^k_{wait} = \begin{cases} T^{k-1}_B, & c^{k-1} > c^k \\ \left(\frac{A}{c^{k-1}}\right) - \left(\frac{A}{c^k}\right) - T^{k-1}_B, & c^{k-1} < c^k \end{cases} \quad (4)$$

where T^i_B is the duration of a single static partition.

In fact, c^k is the slope of the left or right oblique line of the kth operation.

Due to each operation needing to cover the same area of the field, the total duration (T^k_{work}) of the kth operation is calculated as

$$T^k_{work} = A/c^k \quad (5)$$

where A is the area of the field.

Finally, the total operating duration of the kth operation is then calculated by

$$T^k = \sum_2^{k-1} T_{wait}^{k-1} + T_{work}^k. \quad (6)$$

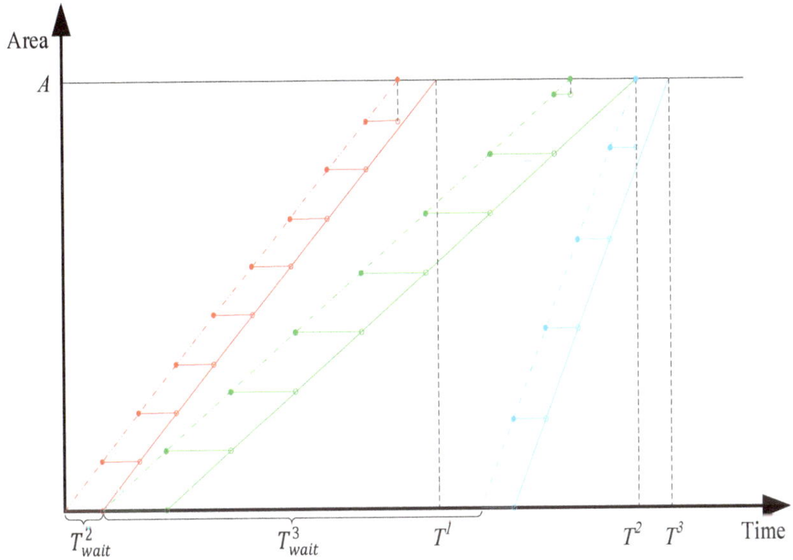

Figure 6. The relationship between the waiting time and the operating capacity per unit time.

2.4. Collaborative Operating System

As shown in Figure 7, the cloud-based coordination system has functionalities to manage the field boundaries, operation tasks, agricultural implements, guidance lines, etc. In this study, only A-B guidance lines were used.

(a)

Figure 7. *Cont.*

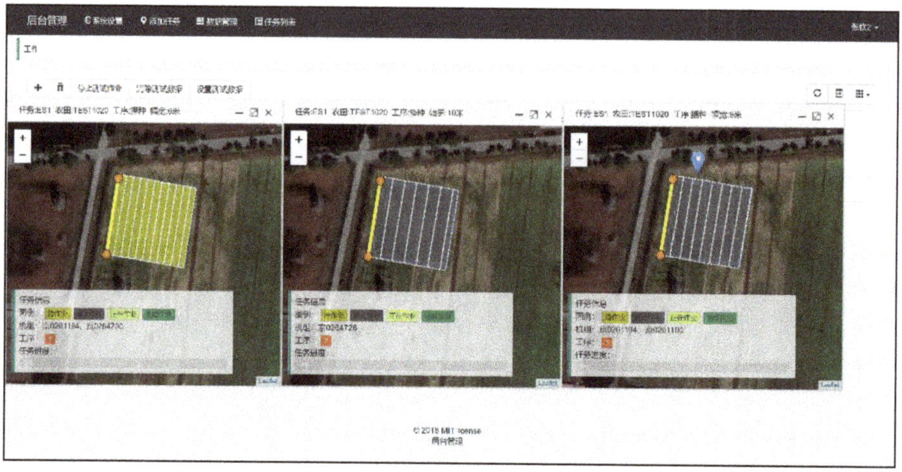

(**b**)

Figure 7. Part of the software interface of the auto-steering-based collaborative operating system for fleet management (FMCOS): (**a**) Software interface for farmland management and (**b**) Software interface for three-operation coordination.

The more important function of the collaborative operating system is to realize a flow-shop working mode with multiple operations and multiple machinery units. Figure 8 shows the module structure of the collaborative operating system, in which the collaborative module for operations can coordinate all the agricultural machinery within the same operation, and the collaborative module for the fleet can coordinate the agricultural machinery between two sequential operations. Obviously, the fleet has $(N-1)$-many pairs of combinations of adjacent operations.

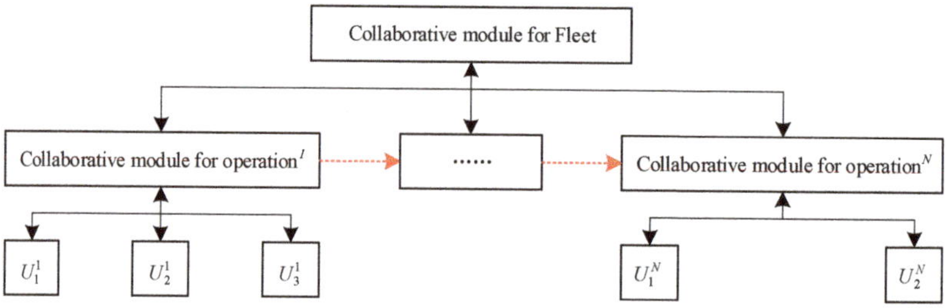

Figure 8. Module structure of the collaborative operating system for fleet management.

2.5. Auto-Steering System

As shown in Figure 9, a hydraulic auto-steering system (HuaCe NX200) was used for testing this system. This auto system has the functionality of sending GNSS positions to and downloading guidance lines from the cloud server. Dual-frequency differential signals with an accuracy of ±1.0 cm from the RTK-GNSS base station located at the experimental field were used.

However, NX200 is a commercial product and cannot be modified. For instance, the guidance line (A-B line) cannot be replaced by the operating strip, and the states also cannot be differentiated with colors. Therefore, a T715C tablet (see Figure 9) with an embedded GPS receiver of a single frequency was used, and an Android-based application was developed for better display. The application can download the position data of the NX200 on the same tractor from the server to replace the T715C's own position data to solve the problem of low accuracy. Although the data transform caused some 2 s delay, the application was able to improve the positioning accuracy of the T715C to ±1.0 cm, which means that the tablet had the same accuracy as the NX200 auto-steering system.

Figure 9. The NX200 auto-steering system for a John Deere 1204 tractor and the Android-based application for a T715C tablet.

3. Experiments

3.1. Simulation Experiments

A simulation model (Matlab 2018a (MathWorks®)) was developed to validate the state-updating algorithm. The modules of the simulation program are shown in Figure 10. In the model, a strip was simulated as the "pushbutton" of UIControl (see Figure 11), while the operation IDs, the strip IDs, and the values of states were stored in the "UserData" field of the control as a three-dimensional array.

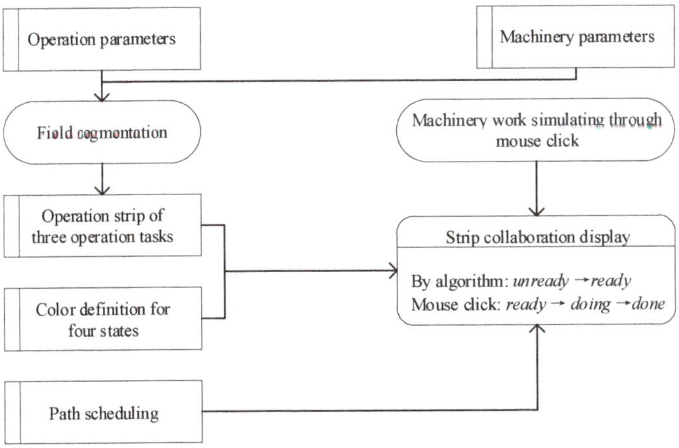

Figure 10. Modules of the simulation model of the collaborative operation.

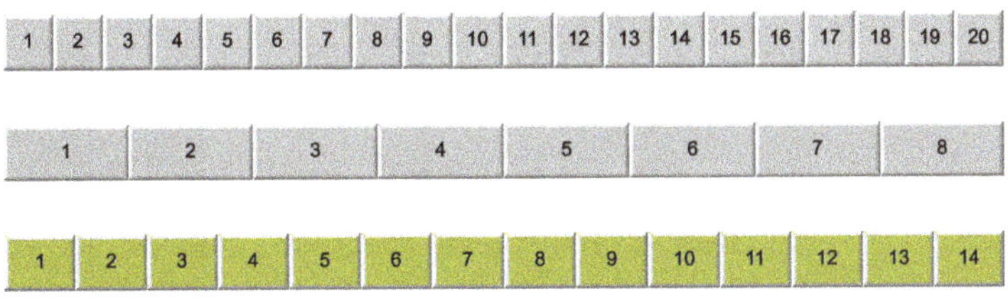

Figure 11. Simulation of the state-updating algorithm using Matlab (MathWorks®).

As shown in Figure 12, in the same operation, the MouseClick command was used to change the state from *ready* to *doing* (simulating the machinery operating on the strip), or from *doing* to *done* (simulating the machinery finishing and leaving the operating strip). When the state of *done* appears, the field operation subsequent to the current field operation will check its *unready* strips to determine the state changes by the algorithm.

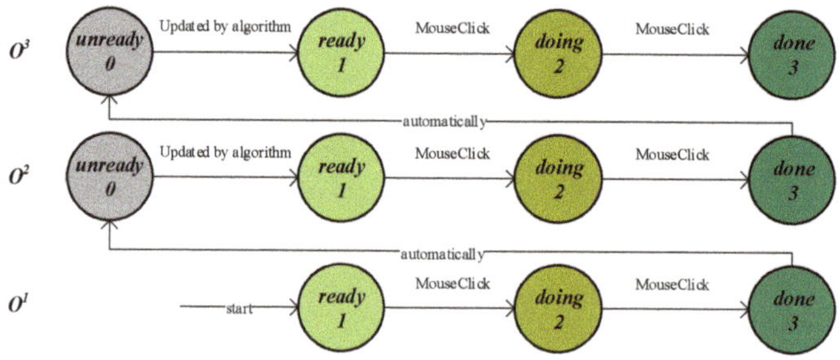

Figure 12. Deterministic finite state automata for the simulation of multi-operation collaboration.

3.2. Field Experiments

Three experimental setups (see Table 2) were designed to test the FMCOS. These setups, ES1, ES2, and ES3, were designed to test the collaboration functions and performance with one operation, two operations, and three operations, respectively. Three sets of NX200 were mounted on three JD1204 tractors (denoted A, B, and C). The superscript of the tractors' symbols (e.g., A^1) represents the sequence of operations.

Table 2. Experimental scheme of FMCOS.

Experiment Scheme	Operating Tasks	Machinery Arrangement *	Width of Strip/m
ES1	One operation	A^1 & B^1	6
ES2	Two operations	A^1 & B^1 VS C^2	6, 8
ES3	Three operations	A^1 VS B^2 VS C^3	8, 5, 6

* In ES1, tractors A and B worked in the first operation. In ES2, tractors A and B worked in the first operation, and then tractor C worked in the second operation. In ES3, operations 1, 2, and 3 were worked by tractors A, B, and C, respectively.

Figure 13 shows the RTK-GNSS base station, which was the experimental field (located at 40.237789° N, 116.581026° E) where all experiments took place, and Figure 14 shows the tractors used for tests.

Figure 13. Real-Time Kinematic–Global Navigation Satellite System (RTK-GNSS) base station for field experiments and the experimental field.

Figure 14. Field experiments scene of (**a**) ES1, (**b**) ES2, and (**c**) ES3.

3.3. Results and Discussion

3.3.1. Simulation Experiments

Figure 15a presents the scenario of two sets of agricultural machinery for the first operation (O^1)—which was worked from the middle of the field to both sides—and one set of agricultural machinery for the second operation (O^2). When clicking on the strip to change the state from *doing* to *ready*, such as $S_6^1 \sim S_{11}^1$, the state of strips related to the second operation, such as S_6^2, changes from *unready* to *ready*, which indicates the appropriate strip for the agricultural machinery of the second operation to work.

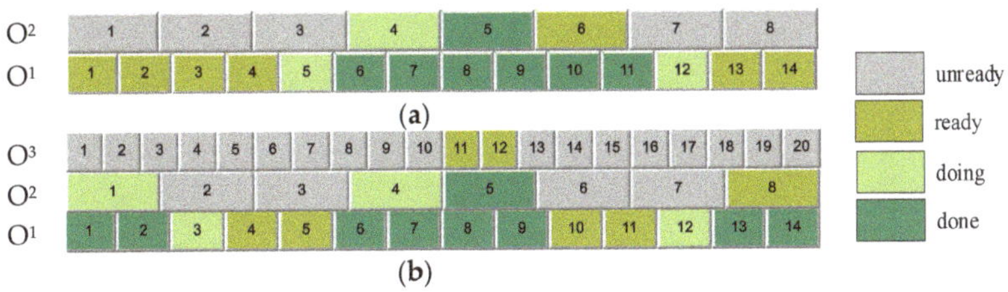

Figure 15. Simulation result in the flow-shop (**a**) Two-operation and (**b**) Three-operation working modes.

Figure 15b simulates the scene of three sets of machinery for the first operation working simultaneously. Two of them worked from the left or right side to the center of the field, while the other worked from the center to the left side of the field. Similarly, the two sets of machinery for the second operation were able to easily find the *ready* strip for working. The machinery for the third operation can choose the appropriate *ready* strips (such as S_{11}^3 and S_{12}^3), and so on.

The results of the above simulations show that the state-updating algorithm can not only well coordinate the state change for multiple machinery within the same operation, but it can also well coordinate two or more operations executed by a fleet of machines.

3.3.2. Field Experiments

Figure 16 depicts screen captures taken during the field experiments of the monitoring interfaces of the collaborative operating system for ES1, ES2, and ES3. ES1 demonstrated the collaborative action of two sets of agricultural machinery (A^1 and B^1) within the same machinery operation (O^1). Figure 16a shows the collaborative interface and Figure 17a shows the trajectory of ES1. From the above referenced pictures, we can see that the tractors could get the same base line from the server and generate the same guidance lines in the NX200 terminal. When the tractor entered or left a strip, the strips of the server and the guidance lines of the terminal changed the state. Therefore, operators were able to choose the *ready* strip to work. ES2 demonstrated the collaborative action of three sets of agricultural machinery (A^1, B^1, and A^2) within two machinery operations (O^1 and O^2). Figure 16b shows the collaborative interface, and Figure 17b shows the trajectory of ES2. The experiment showed that the strip state of the second machinery operation could be updated with the operating progress of the first machinery operation. Therefore, the operators within the same operation are able to choose the *ready* strip for work, and the operators of different operations are also notified in order to select the appropriate strips to enter.

(a) (b)

Figure 16. Monitoring interface of (**a**) ES1 and (**b**) ES2.

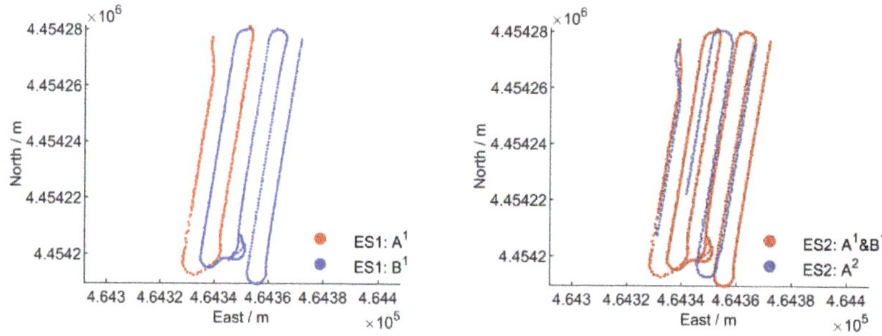

(a) (b)

Figure 17. Trajectory of (**a**) ES1 and (**b**) ES2.

In Figure 17b, in order to show the difference with the second machinery operation, the trajectories of the two tractors (A^1 and B^1) for the same machinery operation were fused.

As shown in Figure 18, ES3 demonstrated the collaborative action of three tractor operations. The experiment process verified that FMCOS could support the coordination of multiple operations. In fact, three-operation collaboration consists of two two-operation collaborations. Therefore, the performance of ES2 has already proved the feasibility and usability of FMCOS.

Figure 18. Monitoring interface of ES3.

The results of the three experiments are shown in Table 3. In ES1, tractors A and B worked in one operation. ES2 consisted of two operations, for which tractors A and B worked in the first operation, then tractor C worked in the second operation. ES3 consisted of three operations, worked by tractors A, B, and C, respectively. Table 3 provides the range of the efficiency improvement for different operating tasks, which have different operation combinations. The maximum efficiency improvement occurs in the experiment of ES3, while the minimum efficiency improvement occurs in the experiment of ES2. In general, the range of efficiency improvement was between 23.05% and 32.72%. It can be concluded that the FMCOS can significantly shorten the waiting time between adjacent operations so as to improve the whole work efficiency.

Table 3. Comparison between Non-FMCOS and FMCOS data in regard to efficiency improvement for ES1, ES2, and ES3.

Experiment	Machinery Arrangement		Operating Time [s]	Non-FMCOS [1] [s]	FMCOS [2] [s]	Efficiency Improvement [3] [%]
ES1	A^1&B^1	A^1	238	951	713	25.03
		B^1	713			
ES2	A^1&B^1		713	1271	978	23.05
	C^2		558			
ES3	A^1		485	1733	1166	32.72
	B^2		685			
	C^3		563			

[1] The sum of operating time; [2] Time from the start to the end of the experiment; [3] $\frac{|(\text{Non-FMCOS})-(\text{FMCOS})|}{\text{Non-FMCOS}} \times 100\%$.

4. Conclusions

A cloud-based collaborative operating system for fleet management (FMCOS) was developed by utilizing a flow-shop working mode and by implementing auto-steering systems. The core technology of the system lies in the state-updating algorithm for the operating strips between adjacent machinery operations and the strategy for adjacent operation connection. The system defined the strip states with four classes: *unready*, *ready*, *doing*, and *done*. The FMCOS was developed after testing the algorithm in a simulation model developed in Matlab (MathWorks®). Three field experiments were carried out, namely, one-operation, two-operation, and three-operation experiments, to verify the functionality and performance of FMCOS. The assessment of the performance of FMCOS showed efficiency improvement up to 25.03% and 32.72% when compared with non-FMCOS. The results of the simulation and actual experiments showed that FMCOS can support in-field fleet management, which can improve the performance of machines in a fleet operating in the same field.

Author Contributions: Conceptualization, C.W. and Z.C.; methodology, C.W. and Z.C.; software, Z.C. and L.Y.; validation, Z.C., D.W., B.S., and Y.L.; formal analysis, Z.C.; investigation, C.W. and Z.C.; resources, C.W.; data curation, Z.C., D.W., and B.S.; writing—original draft preparation, C.W. and Z.C.; visualization, Z.C.; supervision, D.D.B.; project administration, C.W.; funding acquisition, C.W., and Y.L. All authors have read and agreed to the published version of the manuscript.

Funding: This research was funded in part by National Key Research and Development Program of China, grant number 2016YFB0501805. The research was also partly supported by Chinese Universities Scientific Fund, grant number 2018XD003.

Conflicts of Interest: The authors declare no conflict of interest.

References

1. Bsaybes, S.; Quilliot, A.; Wagler, A.K. Fleet management for autonomous vehicles: online PDP under special constraints. *RAIRO Op. Res.* **2019**, *53*, 1007–1031. [CrossRef]
2. Wang, J.; Zhu, Y.; Chen, Z.; Yang, L.; Wu, C. Auto-steering based precise coordination method for in-field multi-operation of farm machinery. *Int. J. Agric. Biol. Eng.* **2018**, *11*, 174–181. [CrossRef]
3. Srensen, C.G.; Bochtis, D.D. Conceptual Model of Fleet Management in Agriculture. *Biosyst. Eng.* **2010**, *105*, 41–50. [CrossRef]
4. Wu, C.C.; Zhou, L.; Wang, J.; Cai, Y.P. Smartphone based precise monitoring method for farm operation. *Int. J. Agric. Biol. Eng.* **2016**, *9*, 111–121. [CrossRef]
5. Noguchi, N.; Will, J.; Reid, J.; Zhang, Q. Development of a master-slave robot system for farm operations. *Comput. Electron. Agric.* **2004**, *44*, 1–19. [CrossRef]
6. Haifeng, L.; Chaoyi, Z. Design and Simulation of a Master-slave Robot for Apple harvest. *Adv. Mater. Res.* **2013**, *816*, 1024–1027.
7. Zhang, X.; Geimer, M.; Noack, P.O.; Grandl, L. Development of an intelligent master-slave system between agricultural vehicles. In Proceedings of the 2010 IEEE Intelligent Vehicles Symposium, San Diego, CA, USA, 24–24 June 2010; pp. 250–255.
8. Zhang, C.; Noguchi, N.; Yang, L. Leader-follower system using two robot tractors to improve work efficiency. *Comput. Electron. Agric.* **2016**, *121*, 269–281. [CrossRef]
9. Zhang, C.; Noguchi, N. Development of Leader-follower System for Field Work. In Proceedings of the 2015 IEEE/SICE International Symposium on System Integration (SII), Nagoya, Japan, 11–13 December 2015; pp. 364–368.
10. Bai, X.; Wang, Z.; Hu, J.; Gao, L.; Xiong, F. Harvester Group Corporative Navigation Method Based on Leader-Follower Structure. *Trans. Chin. Soc. Agric. Mach.* **2017**, *48*, 14–21. [CrossRef]
11. Zhou, K.; Jensen, A.L.; Bochtis, D.D.; Sorensen, C.G. Simulation model for the sequential in-field machinery operations in a potato production system. *Comput. Electron. Agric.* **2015**, *116*, 173–186. [CrossRef]
12. Rodias, E.; Berruto, R.; Busato, P.; Bochtis, D.; Sorensen, C.G.; Zhou, K. Energy Savings from Optimised In-Field Route Planning for Agricultural Machinery. *Sustainability* **2017**, *9*, 1956. [CrossRef]
13. Hameed, I.A.; Bochtis, D.; Sorensen, C.A. An Optimized Field Coverage Planning Approach for Navigation of Agricultural Robots in Fields Involving Obstacle Areas. *Int. J. Adv. Robot. Syst.* **2013**, *10*, 231. [CrossRef]
14. Jensen, M.A.F.; Bochtis, D.; Sørensen, C.G.; Blas, M.R.; Lykkegaard, K.L. In-field and inter-field path planning for agricultural transport units. *Comput. Ind. Eng.* **2012**, *63*, 1054–1061. [CrossRef]
15. Han, X.Z.; Kim, H.J.; Kim, J.Y.; Yi, S.Y.; Moon, H.C.; Kim, J.H.; Kim, Y.J. Path-tracking simulation and field tests for an auto-guidance tillage tractor for a paddy field. *Comput. Electron. Agric.* **2015**, *112*, 161–171. [CrossRef]
16. Bochtis, D.D.; Dogoulis, P.; Busato, P.; Sorensen, C.G.; Berruto, R.; Gemtos, T. A flow-shop problem formulation of biomass handling operations scheduling. *Comput. Electron. Agric.* **2013**, *91*, 49–56. [CrossRef]
17. Senlin, G.; Nakamura, M.; Shikanai, T.; Okazaki, T. Hybrid Petri Nets Modeling for Farm Work Flow. *Comput. Electron. Agric.* **2008**, *62*, 149–158. [CrossRef]
18. Bai, X.; Hu, J.; Gao, L.; Li, T.; Liu, X. Self-tuning model control method for farm machine navigation. *Trans. Chin. Soc. Agric. Mach.* **2015**, *46*, 1–7. [CrossRef]

19. Sadik, A.R.; Urban, B. Flow shop scheduling problem and solution in cooperative roboticscase-study: One cobot in cooperation with one worker. *Future Internet* **2017**, *9*, 48. [CrossRef]
20. Wu, C.; Cai, Y.; Hu, B.; Wang, J. Classification and evaluation of uncertain influence factors for farm machinery service. *Int. J. Agric. Biol. Eng.* **2017**, *10*, 164–174. [CrossRef]
21. Bechar, A.; Vigneault, C. Agricultural robots for field operations. Part 2: Operations and systems. *Biosyst. Eng.* **2017**, *153*, 110–128. [CrossRef]
22. des Dorides, C. *GNSS Market Report*; European GNSS Agency: Luxembourg, 2017.
23. Easterly, D.R.; Adamchuk, V.I.; Kocher, M.F.; Hoy, R.M. Using a vision sensor system for performance testing of satellite-based tractor auto-guidance. *Comput. Electron. Agric.* **2010**, *72*, 107–118. [CrossRef]
24. Cao, R.; Li, S.; Wei, S.; Ji, Y.; Zhang, M.; Li, H. Remote Monitoring Platform for Multi-machine Cooperation Based on Web-GIS. *Trans. Chin. Soc. Agric. Mach.* **2017**, *48*, 52–57. [CrossRef]
25. Deere, J. John Deere Machine Sync. Available online: https://www.deere.com/en_INT/products/equipment/agricultural_management_solutions/guidance_and_machine_control/machine_sync/machine_sync.page? (accessed on 24 November 2014).
26. Jensen, M.F.; Bochtis, D.; Sorensen, C.G. Coverage planning for capacitated field operations, part II: Optimisation. *Biosyst. Eng.* **2015**, *139*, 149–164. [CrossRef]
27. Wu, C.; Yuan, Y.; Han, Y. *Application of Beidou in Agricultural Production Process*; Song, M., Ed.; Publishing House of Electronics Industry: Beijing, China, 2016.
28. Hunt, D. *Farm Power and Machinery Management*, 11th ed.; Waveland Press: Lake County, IL, USA, 2016; pp. 9–17.
29. Zhou, K.; Jensen, A.L.; Bochtis, D.D.; Sorensen, C.G. Quantifying the benefits of alternative fieldwork patterns in a potato cultivation system. *Comput. Electron. Agric.* **2015**, *119*, 228–240. [CrossRef]

MDPI
St. Alban-Anlage 66
4052 Basel
Switzerland
Tel. +41 61 683 77 34
Fax +41 61 302 89 18
www.mdpi.com

Energies Editorial Office
E-mail: energies@mdpi.com
www.mdpi.com/journal/energies

www.ingramcontent.com/pod-product-compliance
Lightning Source LLC
Chambersburg PA
CBHW062050220926
51850CB00010B/2717